SCIENTIFIC REASONING

SCIENTIFIC REASONING: THE BAYESIAN APPROACH

COLIN HOWSON

AND

PETER URBACH

. . . if this [probability] calculus be condemned, then the whole of the sciences must also be condemned.
—Henri Poincaré

Our assent ought to be regulated by the grounds of probability.
—John Locke

Open  Court
La Salle, Illinois

OPEN COURT and the above logo are registered in the U.S. Patent & Trademark Office.

©1989 by Open Court Publishing Company.

First printing 1989.

Library of Congress Cataloging-in-Publication Data
Howson, Colin.
    Scientific reasoning : the Bayesian approach / Colin Howson and
Peter Urbach.
        p.   cm.
    Bibliography: p.
    Includes index.
    ISBN 0-8126-9084-2 : $34.95. ISBN 0-8126-9085-0 (pbk.) : $16.95
    1. Science—Philosophy. 2. Reasoning. 3. Bayesian statistical
decision theory. I. Urbach, Peter. II. Title.
Q175.H87 1988
501—dc19   88-25440
                                                                CIP

CONTENTS

ACKNOWLEDGEMENTS

The authors are very grateful to a number of friends and colleagues who read this book in draft and whose many suggestions led to its substantial improvement. They are John Howard, Martin Knott, Dennis Lindley, and Peter Milne. We are also grateful to Larry Phillips for helpful discussions. Although all of these people would agree with some of what we have written, probably none would agree with it all. Responsibility for the views expressed herein therefore rests entirely with us.

We also express our thanks to Youssef Aliabadi, Helen Brown, Sue Burrett, Alasdair Cameron, Kurt Klappholz, Ginny Watkins, and Gay Woolven for friendly advice, research assistance, and help in preparing the manuscript, to the Suntory-Toyota International Centre for Economics and Related Disciplines for financial assistance, and to the staff of the Open Court Publishing Company for their painstaking editorial work.

Finally, we thank each other. Although we are separately responsible for particular chapters (CH: 2, 3, 9, 11; PU: 1, 4, 5, 6, 7, 8, 10), we have each benefited from regular discussions and the reading and rereading of each other's contributions and this, we believe, has produced a unified exposition of the central Bayesian ideas.

■ PART I

Bayesian Principles

According to the Bayesian view, scientific and indeed much of everyday reasoning is conducted in probabilistic terms. In other words, when evaluating an uncertain claim, one does so by calculating the probability of the claim in the light of given information. Precisely how this is done and why it is reasonable is the topic of this book.

In Part I of the book we shall first introduce the central Bayesian idea, giving some of its intellectual and historical background. This will be Chapter 1. Then in Chapter 2 we shall present the calculus of probability, which constitutes the foundation of the Bayesian approach. This will be done in a relatively formal manner and the question of what it means to say that some hypothesis h has probability $P(h)$ will be considered in Chapter 3. The rest of the book will show how the Bayesian approach gives a penetrating insight into the nature of scientific reasoning far superior to that afforded by any of its rivals.

CHAPTER 1

Introduction

■ a THE PROBLEM OF INDUCTION

Hypotheses usually have a general character relative to the empirical observations they are thought to explain. For instance, Mendel's genetic theory apparently concerns all inherited characteristics in all plants and animals, whereas relatively few of these could ever have been observed. If all our information derives from empirical observation, how can we be sure that any particular explanatory theory is the correct one? This is one version of the traditional problem of induction.

It has, however, sometimes been denied that our stock of information is restricted to empirical observations, a number of philosophers having taken the view that we are also capable of cognizing important synthetic principles which enable the gap between observations and scientific theories to be bridged. Immanuel Kant (1783, p. 9), for example, who claimed that his "dogmatic slumber" had been interrupted by the problem of induction, to which he had been alerted by David Hume's brilliant exposition of it, attempted to provide a principle which was both a priori certain and sufficiently rich to guarantee the truth of the theories of physics. His effort was, however, inadequate. The principle he advocated was just that every event has a cause. Much of Kant's endeavour went into showing that this was an a priori truth, and many of his interpreters have worked hard trying to unravel just what his argument was. But whether valid or not, the principle is irrelevant to the issue at hand, which does not concern whether every event has a cause but asks the very different question: how can one be certain, in any particular case, that one has selected the correct cause of an event out of the huge, indeed infinite, number of possible causes?

Another candidate for a bridging principle between empirical observations and scientific theories is the so-called Principle of the Uniformity of Nature, which Hume (1777, section 32) summed up in the phrase "the future will resemble the past". It is sometimes held that when scientists advocate their theories, they are relying on this principle, at least tacitly.

However, there are two obvious reasons why the theories of science could not be established as definitely true by means of such a principle. First, as it stands, it is empty, for it fails to disclose in what respects the future is supposed to resemble the past. To perform its intended role, the principle would need to be given a specific formulation for application to each case. For example, one such formulation would need to say that, in regard to heated metals, if these have always been observed to expand in the past, then they will do so in the future. It would need a more elaborate formulation to permit the inference that *all* metals would expand if heated, as is usually assumed. But, secondly, as soon as the Uniformity of Nature Principle has been made sufficiently specific for it to connect given observations to particular general laws, its inadequacy as a basis for scientific inference becomes manifest, because its own claim to be accepted as true is now just as questionable as the scientific theory which it was designed to guarantee.

■ b POPPER'S ATTEMPT TO SOLVE THE PROBLEM OF INDUCTION

It would appear then—this is not any longer controversial— that there is no *positive* solution to the problem of induction, that is to say, no solution by whose means particular explanatory theories could be conclusively shown to be true. However, many philosophers and scientists resist the idea, embraced in recent years with particular vigour by Paul Feyerabend, that all theories are on a par and that, for example, standard scientific claims are no better and no worse than those which would commonly be dismissed as the crackpot ideas of a charlatan. Karl Popper, in particular, was concerned to resist such scepticism and put science on a rational footing. He conceded that since scientific theories are never conclusively verifiable, no positive solution exists to the problem of induction. But Popper maintained that theories may, nevertheless, have some worthwhile epistemic status and in some cases be established as epistemically superior to their rivals, this superiority supposedly being an objective feature, independent of anyone's attitude towards them. To this ambitious purpose he pointed out two facts: first, that while theories cannot be logically proved by empirical observations, they can sometimes be refuted by them, and secondly, that their deductive consequences can sometimes be observationally verified. So, for example, 'All swans are white' is refuted by the sighting of a black swan and is, in Popper's terminology, "corroborated" by the observation of a white one. These facts are, of course, not novel discoveries, but what is original with Popper is the attempt to press them into service to show the rationality of science. The attempt, however, is not a success, nor can it be made into one. It is certainly true that by refuting a theory, one can rule out that particular conjecture as the true one. But this, by itself, is practically no help. For, suppose attention were restricted to unrefuted theories, one would still face an infinite class of alternatives, all of which are equally "corroborated", and the problem would remain: how can one choose, from among these, the theory which is most reasonably regarded as the true one?

To illustrate the difficulty, consider the task of discovering some general law governing the coloration of swans. If you were interested in the truth of the matter, you would have to consider many theoretical possibilities. Suppose the total number of swans that will ever exist is n and there are just m different colours, then the number of colour-combinations for the class of swans is m^n. This, then, represents a lower limit for the number of theories concerning the colours swans might have. If we take account of the further possibilities that swans alter their hues from time to time, and from place to place, and that some of them are multicoloured, then it becomes obvious that the number of mutually exclusive rivals to the simple hypothesis 'All swans are white' is immense, indeed it is infinite. To be sure, many of these will have been refuted; for example, the conjecture that all swans are red is falsified by the evidence of white swans. But indefinitely many hypotheses would still remain which are not ruled out in this way by our observations. In fact, theories which specify the colour of every swan at any given instant, which have not been refuted by current observations on the colours of these birds, are corroborated by them. The problem of choosing the best hypothesis amongst these then remains and does not seem to have been advanced by Popper's reflections. (Popper's attempt to solve the induction problem is decisively criticized by Lakatos, 1974, and also by Salmon, 1981.)

■ c SCIENTIFIC METHOD IN PRACTICE

Popper's view that unrefuted but corroborated hypotheses enjoy some special epistemic advantage, independent of anybody's attitude towards them, led him to recommend that scientists ought to seek out such hypotheses. There was also a descriptive aspect to this recommendation, for Popper assumed that mainstream pure science is in fact conducted more or less as he believed it should be. We shall examine this claim.

Two features of scientific reasoning are immediately reflected in Popper's descriptive account. First, it sometimes happens in scientific work that a theory is refuted by some experimental evidence. When this happens, the scientist usually revises the theory or abandons it altogether. (Of course, by revising a theory, one necessarily abandons it altogether! However, when the new theory bears what one intuitively regards as a family resemblance to the old, it is normally spoken of as merely a revision.) Another typical aspect of scientific thought which is taken account of in Popper's scheme is this: when investigating a deterministic theory, scientists draw out some of its consequences, check them by means of a suitable experiment, and if they turn out to be true, often conclude that the theory has been confirmed, or that its claim on our belief in it is strengthened.

Although accurate in these respects, there are two reasons why Popper's descriptive account can have little further success. One has already been touched on in connection with Popper's—in our opinion—unsuccessful attempt to solve the problem of induction. It is that the principles on which he stands are too weak to narrow down the range of alternative hypotheses sufficiently, and, in practice, scientists do have a system of ranking hypotheses by their value and eligibility for serious consideration. As we shall see in Chapter 4, this fact ensures that many typical modes of scientific reasoning are inaccessible to explanation on Popper's principles.

Another limitation of Popper's account is that it focusses attention just on the logical consequences of a theory, whereas most evidence that scientists consider either for or against theories does not come into this category. This arises in a number of ways. First, many deterministic theories that appear in science have no directly checkable deductive consequences and the predictions by which they are tested and confirmed are necessarily drawn only with the assistance of auxiliary theories. Newton's laws are a case in point. These laws concern the forces that operate between masses in general and the mechanical effects of such forces. Observable consequences about particular masses, such as the planets, can be derived only when the laws are combined with hypotheses about the positions and masses of the planets, the mass-distribution of space, and so on. But although such predictions are not direct logical consequences of Newton's theory, that theory is often regarded as being confirmed by them. (Popper did consider this objection, which was pressed with particular vigour by Lakatos, but as we explain in further detail in Chapter 4, his philosophy seems unable to deal with it.)

Secondly, many scientific theories are explicitly probabilistic and, for this reason, have no logical consequences of a verifiable character. An example is Mendel's theory of inheritance. This states the probabilities with which certain combinations of genes occur during reproduction; but, strictly speaking, the theory does not categorically rule out, nor predict, any particular genetic configuration. Nevertheless, Mendel obtained impressive confirmation from the results of his plant-growing trials, results which his theory did not entail but stated to be relatively probable.

Finally, even deterministic hypotheses are frequently confirmed by evidence that is only assigned some probability, for if it is a quantitative theory, its quantitative consequences may need to be checked with imperfect measuring devices, subject to what is known as experimental error. Take as an example a theory which predicts the position of a planet, this prediction being checked using an appropriate kind of telescope. Because of various unpredictable atmospheric conditions affecting the path of light to the telescope as well as other uncontrollable factors, some connected with the experimenter and some with physical vagaries, the actual reading is acknowledged in experimental work not to be completely reliable. For this reason, if the predicted value of an angle were being ascertained, the result of the measurement would normally be reported in the form of a range of values such as $a \pm b$. Here, a is the reading recorded by the instrument, while the interval $a + b$ to $a - b$ signifies the range in which it is judged the true value probably lies. This calculation is usually based on a theory giving the probability that the instrument reading diverges by different amounts from the true value of the measured quantity. Such theories commonly assume that the experimental reading is normally distributed about the true value with a standard deviation of b. (The concepts of a normal distribution and a standard deviation are defined in Chapter 2.) Thus for many deterministic theories, what may appear to be the checking of logical consequences actually involves the examination of experimental effects which are predicted only with a certain probability.

Popper tried to extend the falsificationist idea to the statistical realm but, as we shall show in Chapter 5, there are insuperable difficulties for any such attempt. The eminent statistician R. A. Fisher was also inspired by the idea that evidence may have a decisive negative impact on a statistical hypothesis, akin to its falsification. He called a statistical hypothesis under test the "null hypothesis" and expressed the view that

> the null hypothesis is never proved or established, but is possibly disproved, in the course of experimentation. Every experiment may be said to exist only in order to give the facts a chance of disproving the null hypothesis. (Fisher, 1947, p. 16)

Fisher's theory of significance tests, which prescribes how statistical hypotheses should be tested, has drawn considerable criticism, and several other theories have been advanced in opposition to it. Most notable amongst these is the modified theory of significance testing due to Jerzy Neyman and Egon Pearson. Though they rejected much of Fisher's methodology, their theory owed a good deal to his work, particularly to his technical results. Above all, they retained the idea of bivalent statistical tests in which evidence determines one of only two possible results, that is, the acceptance or rejection of a hypothesis.

■ d PROBABILISTIC INDUCTION: THE BAYESIAN APPROACH

One of the driving forces behind the development of the above-mentioned methodologies was the desire to vanquish, and provide an alternative to, the idea that the theories of science can be and ought to be appraised in terms of their 'probabilities'.

In setting themselves against the ideas of probabilistic induction, Popper and the classical statisticians were opposing a well-entrenched tradition in science and philosophy. Although it has long been appreciated that general scientific theories extend beyond any experimental data and hence cannot be verified (in the sense of being logically entailed) by them, there is, as we have mentioned, a strong tendency in the scientific community, and among philosophers and laymen too, to resist a complete scepticism. Their attitude is that while absolute certainty cannot be expected, nevertheless, the explanations thought up by scientists, and tested by searching experiments, may secure for themselves an epistemic status somewhere between being certainly right and certainly wrong.

This spectrum of degrees of certainty has traditionally been characterised as a spectrum of probabilities. For example, the eminent physicist, mathematician, and philosopher Henri Poincaré reasoned as follows:

> Have we any right, for instance, to enunciate Newton's Law? No doubt numerous observations are in agreement with it, but is not that a simple fact of chance? and how do we know, besides, that this law which has been true for so many generations will not be untrue in the next? To this objection the only answer you can give is: It is very improbable.... From this point of view all the sciences would only be unconscious applications of the calculus of probabilities. And if this calculus be condemned, then the whole of the sciences must also be condemned. (Poincaré, 1905, p. 186)
>
> Thus, in a multitude of circumstances the physicist is often in the same position as the gambler who reckons up his chances. Every time that he reasons by induction, he more or less consciously requires the calculus of probabilities.... (Poincaré, 1905, pp. 183–84)

Similarly, the philosopher and economist, W. S. Jevons:

> Our inferences ... always retain more or less of a hypothetical character, and are so far open to doubt. Only in proportion as our induction approximates to the character of perfect induction, does it approximate to certainty. The amount of uncertainty corresponds to the probability that other objects than those examined, may exist and falsify our inferences; the amount of probability corresponds to the amount of information yielded by our examination; and the theory of probability will be needed to prevent us from over-estimating or under-estimating the knowledge we possess. (Jevons, 1874, vol. 1, p. 263)

Very many scientists have voiced the same idea, that theories have to be judged in relation to their probability in the light of evidence. In fact, as Jon Dorling (1979, p. 180) has observed, it is rare to find any leading scientist writing in, say, the last three hundred years who did not employ notions of probability when advocating his own ideas or reviewing those of others.

A number of philosophers, from James Bernoulli in the seventeenth century to Rudolf Carnap, Harold Jeffreys, Bruno de Finetti, and Frank Ramsey in this century, have attempted to explicate these intuitive notions of inductive probability. There have been two main strands in this programme. The first regards the probabilities of theories as objective, in the sense of being determined by logic alone, independent of our subjective attitudes towards them. The hope was that one would be able to ascertain the probability that a theory is true and thereby place the comparative evaluation of competing explanations on an objective footing. If this could be done, it would provide some kind of solution to the induction problem and establish what might be regarded as a 'rational' basis for science. Unfortunately, as we shall describe later, this approach foundered upon crippling objections and is saved from inconsistency only by arbitrary and highly questionable stipulations.

The other strand of inductive probability treats the probabilities of theories as a property of our attitude towards them; such probabilities are then interpreted, roughly speaking, as measuring degrees of belief. This is called the *subjectivist or personalist interpretation*. The scientific methodology based on this idea is usually referred to as the methodology of *Bayesianism* because of the prominent role it assigns to a famous result of the probability calculus known as Bayes's Theorem.

Bayesianism has experienced a strong revival amongst statisticians and philosophers in recent years, due, in part, to its intrinsic plausibility and also to the weaknesses which have gradually been exposed in the standard methodologies. In the chapters to come, we shall present a detailed account of the Bayesian methodology which we shall show may be applied to gain an understanding of the various aspects of scientific reasoning.

■ e THE OBJECTIVITY IDEAL

The sharpest and most persistent objection to the Bayesian

approach has been that it treats certain subjective factors as relevant to the scientific appraisal of theories. Our reply to this will be that the element of subjectivity admitted in the Bayesian approach is, first of all, minimal and, secondly, exactly right. However, this contradicts an influential school of thought which denies that there should be any subjective element in theory-appraisal at all; such appraisals, according to that school, should be completely objective. Lakatos (1978, vol. 1, p. 1) expressed this objectivist ideal in uncompromising style, thus:

> The *cognitive* value of a theory has nothing to do with its *psychological* influence on people's minds. Belief, commitment, understanding are states of the human mind. But the objective, scientific value of a theory is independent of the human mind which creates it or understands it. Its scientific value depends only on what *objective* support these conjectures have in *facts*. [These characteristic italics are in Lakatos's original mimeographed paper, but were removed in the posthumously edited version.]

It was the ambition of Popper, Lakatos, Fisher, Neyman and Pearson, and others of their schools to develop this idea of a standard of scientific merit which is both objective and compelling and yet non-probabilistic. And it is fair to say that their theories, especially those connected with significance testing and estimation, which comprise the bulk of so-called classical methods of statistical inference, have achieved pre-eminence in the field. The procedures they recommended for the design of experiments and the analysis of data have become the standards of correctness with many scientists.

In the ensuing chapters we shall show that these classical methods are really quite unsuccessful, despite their influence amongst philosophers and scientists, and that their pre-eminence is undeserved. Indeed, we shall argue that the ideal of total objectivity is unattainable and that classical methods, which pose as guardians of that ideal, in fact violate it at every turn; virtually none of those methods can be applied without a generous helping of personal judgment and arbitrary assumption.

■ f THE PLAN OF THE BOOK

The thesis we shall investigate in this book is that scientific reasoning is reasoning in accordance with the calculus of probabilities. In Chapter 2 we shall introduce that calculus, along with the principal theorems that will later serve in an explanatory role with respect to scientific method. We shall then give an account (in Chapter 3) of what we shall mean by the probability of a statement and explain why, quite apart from any considerations of scientific method, it is reasonable to expect that our degrees of belief in hypotheses should obey the axioms of probability.

In Part II, which is also Chapter 4, we shall examine the notion of confirmation and look at the ways in which scientists regard deterministic theories as being confirmed by observations. We shall argue that these characteristic patterns of reasoning are best understood as arguments in probability.

In Part III of the book we turn to statistical hypotheses, where the philosophy of scientific inference has mostly been the preserve of statisticians. Far and away the most influential voice in statistics has been that of the classical statistician, and we shall therefore first give an account of the classical point of view and demonstrate its manifold shortcomings. Thus Chapter 5 deals with Fisher's theory of significance tests, Chapter 6 with his ideas on how causal hypotheses should be tested; Chapter 7 considers the Neyman-Pearson account of significance tests, and Chapter 8 the classical theory of estimation.

In Part IV we present an outline of the Bayesian approach to statistical inference. This will involve first considering the nature of the statistical probabilities that statistical hypotheses purport to be about (are they real or imaginary?). This is done in Chapter 9. Then, in Chapter 10, we show that the difficulties afflicting classical statistical inference are absent from the Bayesian approach, which, we argue, is perfectly satisfactory.

Finally, in Chapter 11, which constitutes Part V of the book, we answer the main criticisms that have been levelled against Bayesianism.

CHAPTER 2

The Probability Calculus

■ a INTRODUCTION

Everything in this book rests upon some property or other of the *probability calculus*. In view of this it will be best to give a brief account of the calculus at the outset and to derive those properties which will be central to later developments. In this way we can avoid breaking up the argument time and again in order to demonstrate that some equation or inequality does indeed follow from basic principles. All the derivations are extremely simple and most require nothing more than elementary algebra.

In this chapter we shall develop the characteristic laws of the calculus purely formally, and then introduce the principal applications of those laws as the book proceeds. However, a glance at the literature on probability will reveal that there are currently *two,* apparently distinct, formal developments of the probability calculus: in one, probabilities are defined on sentences; and in the other, on subsets of some given set (the latter set often being called the space of elementary possibilities). The second, set theoretical development was introduced only in this century, by the Soviet mathematician A. N. Kolmogorov, though the mathematical theory of probability goes back to the seventeenth century. That earlier theory ascribed probabilities directly to hypotheses (to types of sentences or statements), but Kolmogorov's treatment (1933) has now become the standard way of presenting the calculus among statisticians and mathematicians.

The sentence-based presentation tends to be favoured by people who are primarily interested in the application of probability theory to the problems of inductive inference, because there the primary bearers of probabilities are explicitly linguistic entities, namely, hypotheses of one form or other. Since these problems are our primary concern also, we shall adopt that approach too. However, the difference between the two types of development is really just a formal one: the sets in the domain of Kolmogorov's probability function are actually disguised hypotheses, as we shall show presently.

To the uninitiated all this may sound very cryptic. We shall provide a much fuller discussion of these issues later in this chapter, in the course of which we shall also develop the principal consequences of the probability calculus in terms of assignments of probabilities directly to sentences. We shall then introduce, in as informal a way as possible, the notion of a random variable and discuss probability distributions and probability-density distributions over the values of random variables. We shall not be aiming at any great mathematical rigour or purity; there are many texts available on the mathematics of probability, and where readers think they will benefit by more information on any of these topics, they are invited to consult any or all of these texts. We want only to introduce as much technical material as is necessary to understand what is going on, and so there will be few proofs, and many if not all mathematical corners will be cheerfully cut.

■ b SOME LOGICAL PRELIMINARIES

We shall be employing some, but not many, of the notions of elementary logic—principally just the so-called logical connectives, or truth-functional operations, 'and', 'or', and 'not'. We shall use a fairly standard notation here, symbolising 'and' by '&', 'or' by '∨', and 'not' by '∼'. Thus $a \& b$ is the conjunction of the sentences a and b, and it will be taken to be true just when a and b are both true (we should strictly put '$a \& b$' in quotation marks, but the text would look awful and where no confusion is likely to arise we shall omit them). '∨' is inclusive 'or', that is, $a \vee b$ will be false just when a and b are both false. Every sentence will be taken to be true or false when all its referring terms are assigned a specific reference (this is called the condition of bivalence). $\sim a$, of course, will be true just when a is false. Occasionally we shall make use of the biconditional '↔': $a \leftrightarrow b$ is true just in case a and b are both true or both false.

We shall use the notation $a \vdash b$ to signify that a entails b deductively; and to say that a entails b deductively is simply to say that it is impossible, independently of the state of the world, for a to be true and b false. $a \vdash a$, $a \vdash \sim\sim a$, $a \& b \vdash a$,

$a \vdash a \lor b$, are some simple examples of entailment.

The notation $a <=> b$ will signify that a is equivalent to b; that is to say, it is impossible, independently of the state of the world, for a and b to possess different truth values (i.e. *true, false*). $a <=> a$, $a <=> \sim \sim a$, $a \& b <=> b \& a$, $a \lor b <=> b \lor a$ are some simple examples of equivalence. It is not difficult to infer that $a <=> b$ just in case $a \vdash b$ and $b \vdash a$, and that $a <=> b$ just in case $a \leftrightarrow b$ is a tautology.

A *tautology* is a sentence, like "if it is not the case that it is not raining here now, then it is raining here now", which is true independently of the state of the world, and a *contradiction* is a statement, like "Socrates is a man and it is not the case that Socrates is a man", which is false independently of the state of the world. Both types of statement are easy to generate using the connectives: thus, $a \lor \sim a$, $\sim(a \& \sim a)$ are examples of tautologies so generated, and it is fairly obvious that once one has a tautology, one can obtain a contradiction (or vice versa) by simply negating it. Moreover, it is also a simple inference from the definitions of tautology and contradiction that any statement deductively entails a tautology and is entailed by a contradiction.

Our notions of entailment and equivalence are stronger than the purely logical ones, as they are to be understood as incorporating all of contemporary mathematics. Thus, for example, if x is any individual, and A and B any two sets, then '$x \in A \& x \in B$' will be regarded as deductively entailing '$x \in A \cap B$'. $\in$ as usual signifies the membership relation, and $\cap$ the intersection of the two sets A and B, that is to say, the set whose members are common to A and B. The union $A \cup B$ of two sets A and B is the set whose members are in A or in B or in both A and B. The complement $B - A$ of a set A with respect to some set B is the set whose members are the members of B excluding all those which are also members of A. '$A \subseteq B$' signifies that A is a subset of B; that is, every member of A is also a member of B. It follows immediately that every set is a subset of itself. A *singleton* set is a set with one member only. The empty set is, as usual, denoted by the symbol $\emptyset$.

■ c THE PROBABILITY CALCULUS

c.1 The Axioms

Let us assume that we are given a class S of sentences $a, b, c, \ldots$, which may also contain conjunctions, disjunctions, and negations of any given sentences which it contains. At the extreme, S may be *closed* under these truth-functional operations; that is to say, S may be such that it contains a & b, a **v** b, $\sim a$, and $\sim b$, whenever it contains a and b. We shall not assume this to be the case, though we shall assume that S is non-empty, and also that it contains at least one tautology. A probability function on S is a function which assigns non-negative real numbers to the sentences in S, in such a way that every tautology is assigned the value 1, and the sum of the probabilities of two mutually inconsistent sentences is equal to the probability of their disjunction; in other words, the following three conditions, or axioms, are satisfied:

(1) $P(a) \geq 0$ for all a in S
(2) $P(t) = 1$ if t is a tautology
(3) $P(a \textbf{ v } b) = P(a) + P(b)$ if a and b and $a \textbf{ v } b$ are all in S, and a and b are mutually inconsistent; i.e., such that one entails the negation of the other.

These three conditions suffice to generate that part of the probability calculus dealing with so-called *absolute,* or *unconditional, probabilities.* (3) is often called the *Additivity Principle,* since it states that P adds over disjunctions of pairs of mutually inconsistent statements. As we shall show shortly, (3) together with the other axioms implies that P adds over all finite disjunctions of mutually exclusive statements.

So-called *conditional probabilities* are given as a function $P(\cdot \mid \cdot)$ of two variables, called the conditional probability function based on P, which satisfies the condition

(4) $P(a \mid b) = \dfrac{P(a \, \& \, b)}{P(b)}$,

where a, b, and a & b are in S, and where $P(b) \neq 0$. Many authors take $P(a \mid b)$ to be defined by this condition; we prefer to regard (4), however, as a postulate on a par with (1)–(3). (This means that '$P(a \mid b)$' is in effect a primitive of the theory in the same way as $P(a)$.) The reason for this is that in some interpretations of the calculus, independent meanings are given conditional and unconditional probabilities, and equation (4) becomes a synthetic, not an analytic, truth.

In what follows, any result involving $P(a \mid b)$, for any a,b, will be taken to have satisfied the conditions stated in (4), that $P(b) > 0$, and that the statements b and a & b are in S.

c.2 The Set-Theoretic Approach: Kolmogorov's Axioms

We said that Kolmogorov defined his probability function, which we shall label P', as a function on sets, not sentences—on a field F of subsets of some set U, to be precise. To say that F is a field of subsets of U is to say that U is in F, and that F contains all complements of sets in F relative to U, and also that F contains all unions and intersections of sets in F (more briefly, F is closed under complements, unions, and intersections). Kolmogorov's axioms then state that

(1)' $P'(A) \geq 0$, for all A in F

(2)' $P'(U) = 1$

(3)' $P'(A \cup B) = P'(A) + P'(B)$, if $A \cap B = \emptyset$

(4)' $P'(A \mid B) = \dfrac{P'(A \cap B)}{P'(B)}$, where $P'(B) \neq 0$

Kolmogorov included a further axiom, that of continuity, and a further condition on the field F, both of which we shall discuss shortly. U is intended by Kolmogorov (1933, p. 4; he uses the letter S) to represent a set of mutually exclusive and exhaustive outcomes, often called the outcome space, of some repeatable experiment E. F, the domain of P', is a set of types of outcome, represented extensionally, of E. For example, if the experiment is throwing a die, then U would usually (but not necessarily) be the set $\{1, \ldots, 6\}$, and F the set of all subsets of U; and $\{2,4,6\}$ would then represent the type of outcome 'an even number is shown on the uppermost face'. F does not necessarily contain the singleton members of U, nor is it always (and sometimes, for mathematical reasons, it cannot be) the set of *all* subsets of U.

c.3 Sets or Sentences?

Defining probabilities on sets rather than on sentences can be regarded as only a notational difference between the two formulations. To see this, think of any given sentence as representing all those possible states of affairs consistent with its truth. We can, in other words, represent a sentence by means of its *extension,* that is to say, the set of all logical possibilities left open by a. These possibilities will be required to be the possibilities formulable within some specified system of categories, or language-system, otherwise they would be limitless and the set of all possibilities would correspondingly be ill-defined. For example, suppose that the language system is a very simple one, sufficient for discussing whether the individuals in some domain of discourse possess a given property, A say, but nothing more. Suppose these individuals are named b_1, b_2, b_3, and so on, up to b_n, for some n. Then if these n individuals exhaust the domain of discourse, and they are all distinct, there are 2^n ways in which the property A may be distributed among them. Thus if a is the sentence 'a_1 has the property A', then a is clearly true in just 2^{n-1} of these, and this latter set is therefore the subset of all the elementary possibilities describable within this system which are admitted by a. Of course, such a language is admittedly extremely simple, and far too simple for most descriptive purposes, but its very simplicity makes it useful to illustrate a point. We shall use its services again in the next chapter, where its simple structure will allow us to gain some insights into so-called logical probability measures.

Any intermediate text on modern logic will reveal that there are several ways of making 'the set of possibilities admitted by a', where a is a sentence from some well-defined formal language, into a precisely characterised mathematical object. We shall not pursue these mathematical details further: the notion is intuitively clear enough. Let us represent by $M(a)$ the set of possibilities admitted by a. It is not difficult to see that logical operations on and relations between the sentences a are reflected in corresponding set-theoretic operations on and relations between the $M(a)$: the two structures, that of the sentences under the logical operations and relations, and that of the $M(a)$ under the corresponding set-theoretic operations and relations, are said to be homomorphic. If we were to identify all logically equivalent sentences, then the structures are isomorphic, since the mapping associating a with $M(a)$ then becomes one-to-one, or bijective; in the present case, the mapping is merely many-one, and therefore only a homomorphism. Homomorphism means 'having a similar structure', and this similarity of structure between the sentences a and the sets $M(a)$ is manifested in the following identities:

$$M(a \lor b) = M(a) \cup M(b)$$

$$M(a \ \& \ b) = M(a) \cap M(b)$$

$$M(\sim a) = V - M(a),$$

where V is the universal set (universal relative to the language) of possibilities admitted by a tautology. Thus, where f is a contradiction, $M(f) = \emptyset$, the empty set. The condition that a and b are mutually inconsistent in axiom 3 of the probability calculus translates into the set-theoretic condition $M(a) \cap M(b) = \emptyset$. That $a \vdash b$, meaning that a deductively entails b, which is to say that it is impossible for a to be true and b false, translates into the condition that $M(a) \subseteq M(b)$, and we therefore have a simple set-theoretic analogue of deducibility. Logical equivalence has an even simpler translation. For $a <=> b$ if and only if $a \vdash b$ and $b \vdash a$, that is, if and only if $M(a) \subseteq M(b)$ and $M(b) \subseteq M(a)$, which is true if and only if $M(a) = M(b)$.

If we now interpret the sets A in the domain $\mathbf{F}$ of Kolmogorov's probability function as $M(a)s$, the a's taken from some well-defined language L, and replace the resulting terms $P'(M(a))$ by $P(a)$, setting U equal to V, it is easy to see that Kolmogorov's axioms 1'–4' translate under this identification into our axioms 1–4. Thus, for the die-throwing experiment E we mentioned in the discussion of Kolmogorov's own interpretation of the sets U and $\mathbf{F}$, the sentences a would be drawn from a language whose elementary, or atomic, sentences are of the form 'the number i is on the uppermost face', for $i = 1, \ldots, 6$. Where a_i is the sentence quoted, we can take $M(a_i)$ to be $\{i\}$, the singleton containing i. If $\mathbf{F}$ is the set of all subsets of $\{1, \ldots, 6\}$, then each set in $\mathbf{F}$ is the extension of some sentence obtained by purely truth-functional operations on these atomic sentences. In other words, we can, and Kolmogorov himself implicitly tells us we should, regard the sets in $\mathbf{F}$ as—essentially—descriptions, in this case descriptions of generic events defined in the outcome space U of E.

We shall, in section **d** below, say a little more about event-describing sentences, but before doing so we mention a further important feature of Kolmogorov's presentation (but one that is not crucial to an understanding of the main arguments of the book—section **c.4** can be skipped by the reader impatient of detail).

c.4 Countable Additivity

Kolmogorov required—for reasons largely of mathematical simplicity—closure of the domain of P' under the operations of union, intersection, and complement relative to U. In addition to closure under these finite operations, however, he stipulated that $\mathbf{F}$ be closed also under denumerably *infinite* unions and intersections. ((A_i) is a denumerably infinite family of sets if the index i ranges over all the positive integers.) Any field obeying this further closure condition is called a sigma field of sets.

These infinite unions and intersections, were the A_i regarded as being $M(a_i)$ for a corresponding family of sentences a_i, would correspond to infinite disjunctions and conjunctions indexed by the positive integers, like $a_1 \lor a_2 \lor a_3 \lor \ldots$, and $a_1 \mathbin{\&} a_2 \mathbin{\&} a_3 \mathbin{\&} \ldots$. Clearly, nobody can actually write out an infinite disjunction or conjunction, but standard languages have a way round this in some cases. Thus, the true statement 'some positive whole number is even and prime' clearly represents what would be an infinite disjunction if we could write it all out: 'one is even and prime or two is even and prime or ...'. Another true statement about positive integers is 'every prime greater than two is odd', and this represents the infinite 'conjunction' 'three is odd and five is odd and seven is odd and ...'. These infinite operations are, however, specifically allowed for in set theory, even for arbitrary collections of sets, and so one can develop the formal theory of probability in the Kolmogorovian form without having to worry about whether one is always making sense at the informal linguistic level. At the purely mathematical level one is, and so a smooth formal development becomes much easier—which is another reason why mathematicians largely choose to work within that framework. Adding corresponding explicit infinitary operations to the standard (first-order) logical languages results, incidentally, in a system which shares many features of first-order ones: its class of valid sentences is axiomatisable, for example.

Having stipulated that the domain of P', that is $\mathbf{F}$, be a sigma field, Kolmogorov adopted as an additional axiom over and above $(1)'-(4)'$, his so-called Axiom of Continuity. This is equivalent to requiring that where UA_i is any denumerable union of mutually disjoint sets in $\mathbf{F}$, then $P'(UA_i)$ is equal to $\sum_i P'(A_i)$. (There is no need to stipulate that the infinite sum exists: this follows from the fact that the sequence of partial, finite sums is a bounded monotone sequence.) This infinite-additivity condition is more often called the *Principle of Countable Additivity,* and it is very much stronger than any corresponding principle of finite additivity (we shall see shortly that it follows from the other axioms that probabilities add over any finite set of mutually exclusive statements), as can be seen in the context of the following simple example. Suppose that we are considering the 'experiment' of selecting a positive integer in some way. If we adopt the Principle of Countable Additivity, then it is impossible for each integer to have the same probability of being selected, for the probability of selecting *some* integer is one, which is equal to the probability of selecting one or two or three or . . . , which is equal, given the infinite additivity condition, to the probability of selecting one plus the probability of selecting two plus the probability of selecting three, plus . . . , and so on. But it is obvious that if these summands were equal and positive, the sum would diverge, while if they were equal and zero, their sum would still be zero in the limit. If the probabilities are only finitely additive, on the other hand, it is not difficult to see that it is consistent to make the probability of each integer being selected the same for all integers, namely 0.

We shall not list the Principle of Countable Additivity among the fundamental axioms of the probability calculus, because while it is a legitimate constraint in one of the two principal interpretations of the calculus we shall be looking at, it is certainly not so in the other. Chapter 3 and Chapter 9 will discuss at length these two interpretations, but some preliminary discussion will be helpful at this point.

■ d TWO DIFFERENT INTERPRETATIONS OF THE PROBABILITY CALCULUS

We have presented the fundamental principles of the probability calculus in a rather unmotivated and abstract way because, as has been remarked since the beginning of the nineteenth century, there are at least two quite distinct notions of probability, both of which appear to satisfy the formal conditions 1–4 above. According to one of these, the probability calculus expresses the fundamental laws regulating the assignment of objective physical probabilities to events defined in the outcome spaces of stochastic experiments (a classical example of a stochastic trial, and, because of its simplicity, one we shall make much use of subsequently, is that of tossing a coin and noting which face falls uppermost).

The other notion of probability is epistemic. This type of probability is, to use Laplace's famous words, "relative in part to [our] ignorance, in part to [our] knowledge" (1820, p. 6): it expresses numerically degrees of uncertainty in the light of data. We shall be discussing these two notions in considerable detail in the following chapters; we mention them here not only because they involve distinct interpretations of the probability-values themselves, but also because the statements to which they assign probabilities are of quite distinct types. In the latter, epistemic, interpretation, the statements to which the probabilities are assigned are specific hypotheses, like 'the Labour Party will not win the next General Election in the UK'. As we shall see, however, there is more than one epistemic interpretation, an ostensibly person-independent one, and a frankly subjective one.

There is also more than one objectivist interpretation of the probability function, and in at least one of these, the statements describe *generic* events which can arise as possible outcomes of a stochastic trial or experiment. But here we are faced with an apparent difficulty: 'the coin lands heads' is true or false relative to specific tosses of specific coins. How can a sentence describe the generic event of landing heads? The answer is, in brief, that it does so by leaving the referents of the appropriate singular terms in the sentence unspecified within the type, or class, from which they come. In a natural language such as English, we are not accustomed to the notion of a syntactically well-formed but partially uninterpreted sentence. Within the notation of formal logic, however, the notion is easily characterised. Thus, $B(a)$, where a is an individual name, or constant, and B a predicate symbol, describes a *specific* individual event when a and B are both fixed (a might, for example, be made to refer to the next toss of this coin, and B be the predicate, *lands heads*). The same formal sentence $B(a)$ will be said to describe the *generic* event of this coin's landing heads, when a is not specified as any one of the tosses of this coin, but B remains fixed as the predicate *lands heads,* referred to the class of tosses of this coin. It might seem more appropriate to employ a free-variable formula $B(x)$ to refer to the generic event, and some authors do. Nothing is wrong with this in principle, but choosing that expression would deny us the use of ordinary vernacular sentences, where there is no syntactical distinction between terms which have definite as opposed to indefinite reference. As probabilities are characteristically assigned to vernacular sentences and not to the formulas of formal languages, we shall accordingly use one and the same sentence for both specific and generic reference, distinguishing those uses by appropriate contextual stipulation.

The reader should note that even a sentence like 'This coin lands heads on the ith toss' is as ambiguous between specific and generic reference as 'this coin lands heads'. The term 'the ith toss' refers implicitly to a finite or infinite sequence of tosses of this coin, but again, we may choose to make that reference generic or specific. The motive, speaking for objective-probability theorists of a certain stripe, for attaching probabilities to generic events, or rather to the sentences characterising those generic events, is, as we shall see in Chapter 9, that the associated probability numbers are not intended to describe features of the outcome of any particular performance of the experiment, but, on the contrary, to express the frequency with which the event in question occurs in long sequences of performances of that experiment. But some people have also tried to construct theories of objective probability in which these probabilities are attached to predictions about the outcome of a specific performance of some stochastic trial. We shall defer all discussion of these attempts to the appropriate chapter, however, and proceed now to derive the familiar 'laws' of the probability calculus from the axioms 1–4.

■ e USEFUL THEOREMS OF THE CALCULUS

In deriving these consequences of the calculus, some standard implications and identities of propositional logic are used, and the reader unfamiliar with these might find it helpful to check their validity by considering the corresponding set-theoretic counterparts under the mapping a to $M(a)$, described above.

The first result states the well-known fact that the probability of a sentence and that of its negation sum to 1:

(5) $P(\sim a) = 1 - P(a)$

Proof.

$a \vdash \sim\sim a$. Hence by (3) $P(a \lor \sim a) = P(a) + P(\sim a)$. But by (2) $P(a \lor \sim a) = 1$, whence (5).

Next, it is simple to show that contradictions have zero probability:

(6) $P(f) = 0$, where f is any contradiction.

Proof.

$\sim f$ is a tautology. Hence $P(\sim f) = 1$ and by (5) $P(f) = 0$.

Our next result states that equivalent sentences have the same probability:

(7) If $a \Leftrightarrow b$ then $P(a) = P(b)$.

Proof.

First, note that $a \lor \sim b$ is a tautology if $a \Leftrightarrow b$. Assume that $a \Leftrightarrow b$. Then $P(a \lor \sim b) = 1$. Also if $a \Leftrightarrow b$, then $a \vdash \sim \sim b$; so $P(a \lor \sim b) = P(a) + P(\sim b)$. But by (5) $P(\sim b) = 1 - P(b)$, whence $P(a) = P(b)$.

We can now prove the important property of probability functions that they respect the entailment relation; to be precise, the probability of any consequence of a is at least as great as that of a itself:

(8) If $a \vdash b$ then $P(a) \le P(b)$.

Proof.

If $a \vdash b$ then $[a \lor (b \ \& \sim a)] \Leftrightarrow b$. Hence by *(7)* $P(b) = P(a \lor (b \ \& \sim a))$. But $a \vdash \sim(b \ \& \sim a)$ and so $P(a \lor (b \ \& \sim a)) = P(a) + P(b \ \& \sim a)$. Hence $P(b) = P(a) + P(b \ \& \sim a)$. But by (1) $P(b \ \& \sim a) \ge 0$, and so $P(a) \le P(b)$.

From (8) it follows that probabilities are numbers between 0 and 1 inclusive:

(9) $0 \le P(a) \le 1$, for all a in S.

Proof.

$f \vdash a \vdash t$, where f is any contradiction and t any tautology. Hence by (6), (2), and (8): $0 \le P(a) \le 1$.

We shall now demonstrate the general (finite) additivity condition:

(10) Suppose $a_i \vdash \sim a_j$, where $1 \le i < j \le n$. Then $P(a_1 \lor \ldots \lor a_n) = P(a_1) + \ldots + P(a_n)$.

Proof.

$P(a_1 \lor \ldots \lor a_n) = P((a_1 \lor \ldots \lor a_{n-1}) \lor a_n)$, assuming that $n > 1$; if not the result is obviously trivial. But since $a_i \vdash \sim a_j$, for all $i \ne j$, it follows that $(a_1 \lor \ldots \lor a_{n-1}) \vdash \sim a_n$, and hence $P(a_1 \lor \ldots \lor a_n) = P(a_1 \lor \ldots \lor a_{n-1}) + P(a_n)$. Now simply repeat this for the remaining $a_1, \ldots, a_{n-1}$ and we have (10). (This is essentially a proof by mathematical induction.)

Corollary. If $a_1 \mathbf{v} \ldots \mathbf{v} a_n$ is a tautology, and $a_i \vdash \sim a_j$ for $i \neq j$, then $1 = P(a_1) + \ldots + P(a_n)$.

Our next result is often called the 'theorem of total probability'.

(11) If $a_1 \mathbf{v} \ldots \mathbf{v} a_n$ is a tautology, and $a_i \vdash \sim a_j$ for $i \neq j$, then $P(b) = P(b \ \& \ a_1) + \ldots + P(b \ \& \ a_n)$, for any sentence b.

The proof is left to the reader.

A useful consequence of this is the following:

(12) If $a_1 \mathbf{v} \ldots \mathbf{v} a_n$ is a tautology and $a_i \vdash \sim a_j$ for $i \neq j$, and $P(a_i) > 0$, then for any sentence b
$P(b) = P(b \mid a_1)P(a_1) + \ldots + P(b \mid a_n)P(a_n)$.

Proof.

A direct application of (4) to (11).

Corollary. $P(b) = P(b \mid c)P(c) + P(b \mid \sim c)P(\sim c)$, for any c.

We shall now develop some of the important properties of the function $P(a \mid b)$. Recall that we are assuming in these derivations that the second argument b of $P(a \mid b)$ has positive probability (though this of course is not in practice always going to be the case).

(13) Let b be some fixed sentence, and define the function $Q(a)$ of the one variable sentence a to be $P(a \mid b)$. Then $Q(a)$ satisfies axioms 1–3, that is, it is a probability function.
Now define 'a is a tautology modulo b' simply to mean '$b \vdash a$' (for then $b \vdash (t \leftrightarrow a)$, where t is a tautology, so that relative to b, a and t are equivalent), and 'a and c are exclusive modulo b' to mean '$b \ \& \ a \vdash \sim c$'; then

(14) $Q(a) = 1$ if a is a tautology modulo b; and the corollary

(15) $Q(b) = 1$;

(16) $Q(a \mathbf{v} c) = Q(a) + Q(c)$, if a and c are exclusive modulo b.

The proofs of (13)–(16) are very straightforward and are left for the reader.

We are now in a position to state the results which variously go under the name Bayes's Theorem. This theorem, or rather these theorems, are named after the eighteenth century English clergyman Thomas Bayes. Although Bayes, in a posthumously published and justly celebrated Memoir to the Royal Society of London (1763), derived the first form of the theorem named after him, the second is due to the great French mathematician Laplace.

Bayes's Theorem (first form)

(17) $P(a \mid b) = \dfrac{P(b \mid a) \, P(a)}{P(b)}$, where $P(a), P(b) > 0$.

Proof.

$$P(a \mid b) = \frac{P(a \, \& \, b)}{P(b)} = \frac{P(b \mid a) \, P(a)}{P(b)}.$$

This result which, as we have seen, is mathematically trivial, is nonetheless of central importance in the context of so-called Bayesian inference; there a is usually the hypothesis to be evaluated relative to empirical data b, and this form of Bayes's Theorem thus states that the probability of the hypothesis conditional on the data (or the *posterior probability* of the hypothesis) is equal to the probability of the data conditional on the hypothesis (or the *likelihood* of the hypothesis) times the probability (the so-called *prior probability*) of the hypothesis, all divided by the probability of the data.

Bayes's Theorem (second form)

(18) If $P(b_1 \lor \ldots \lor b_n) = 1$ and $b_i \vdash {\sim}b_j$ for $i \neq j$, and $P(b_i), P(a) > 0$ then

$$P(b_k \mid a) = \frac{P(a \mid b_k) \, P(b_k)}{\displaystyle\sum_{i=1}^{n} P(a \mid b_i) \, P(b_i)}$$

The proof is straightforward, and is left to the reader.

(19) (Corollary) If $b_1 \lor \ldots \lor b_n$ is a tautology, then if $P(b_i) > 0$ and $b_i \vdash {\sim}b_j$ and $P(a) > 0$, then

$$P(b_k \mid a) = \frac{P(a \mid b_k) \, P(b_k)}{\sum\limits_{i=1}^{n} P(a \mid b_i) \, P(b_i)}$$

■ f RANDOM VARIABLES, DISTRIBUTION FUNCTIONS, AND PROBABILITY DENSITIES

f.1 Random Variables

Bayes's Theorem or Theorems can be expressed not only as results about probabilities, but also as results about probability densities. In order to explain what probability densities are, we must first introduce the notion of a random variable. A random variable is a quantity X capable, depending on the actual state of affairs, of taking on different numerical values, and having a definite probability of being found in any open or closed interval of real numbers—that is to say, since we are assigning probabilities to statements, the statement that X lies in any such interval is assigned a definite probability.

Since it is a quantity which may take different values relative to different possible states of affairs, a random variable is formally a *function* whose domain is the set of those distinct possibilities and whose values are real numbers. Thus, in terms of the notation we used in section **c.3** of this chapter, we can write $M(x_1 < X < x_2)$ as $\{w : x_1 < X(w) < x_2\}$, meaning, the set of all w such that $x_1 < X < x_2$, where the possibilities w are the members of $M(t)$, for any tautology t, and $X(w)$ is the value taken by X relative to w.

The set of possibilities w on which any given random variable is defined is identified, in applications in statistics and mathematical physics, with a set of mutually exclusive and exhaustive outcomes of some repeatable experimental procedure. It is this set, called the outcome space of the experiment, that Kolmogorov's U represents (above, section **c.2**). The outcome space is often taken to be a set of numbers: for example, we might be measuring the number of decay particles emitted per unit time by some radioactive element, or the heights of some selected group of adult Americans, or (at a more familiar and unsophisticated level) the number of heads obtained in n tosses of some coin, and record the outcomes simply by means of these numbers. The random variables in these cases are then the identity functions on these sets, and in practice we usually call these numbers themselves the random variables. This confusion of random variables (functions) with numbers (scalars) is harmless in most cases, but there are occasions where the confusion can lead to paradox (see Chapter 11).

Random variables are a very convenient method of describing experimental outcomes, even when those outcomes appear not to be at all numerical in nature. Often, for example, it is useful to describe the outcome of a single toss of a coin in terms of a binomial, or two-valued, random variable: 0 if it is a head, 1 if it is a tail, or vice versa (clearly, the actual numbers 0 and 1 have no significance in themselves; any other pair, like 1111 and 35, could be chosen). In general, binomial random variables are used for the description of outcomes of an experiment in which merely the presence or absence of some particular character C is all the experimenter is interested in. In such cases we set $X(w) = 1$ if w has C, and $X(w) = 0$ if not; in this way the set of possible outcomes of the experiment is collapsed by X into just two. Thus in such cases X in effect partitions the set of outcomes of the experiment into two: that subset on which X takes the value 1, and that subset on which X takes the value 0. Clearly, by means of appropriately chosen random variables defined on a set S of outcomes of some experiment, we can partition S in any way we wish, and into as many non-overlapping cells, up to the limit set by the original number of possibilities w as we wish.

Statements of the form $X < x'$ play a fundamental role in mathematical statistics. Clearly, the probability of any such statement will vary with the choice of the real number x; it follows that this probability is a function $F(x)$, the so-called *distribution function,* of the real variable x. Thus, where P is the probability measure concerned, the value of $F(x)$ is defined to be equal, for all x, to $P(X < x)$ (although F depends therefore also on X and P, these are normally apparent from the context and F is usually written as a function of x only). Some immediate consequences of the definition of $F(x)$ are that

(i) $F(-\infty) = 0$, and $F(+\infty) = 1$
(ii) if $x_1 < x_2$ then $F(x_1) \leq F(x_2)$, and
(iii) $P(x_1 \leq X < x_2) = F(x_2) - F(x_1)$.

Distribution functions are not necessarily functions of one variable only. For example, we might wish to describe a possible eventuality in terms of the values taken by a number of random variables. For example, consider the 'experiment' which consists in noting the heights (*X*, say) and weights *(Y)* jointly of members of some human population. It is usually accepted as a fact that there is a joint (objective) probability distribution for the vector variable *(X,Y)*, meaning that there is a probability distribution function $F(x,y) = P(X < x \ \& \ Y < y)$. Mathematically this situation is straightforwardly generalised to distribution functions of *n* variables. For the sake of simplicity we shall try as much as possible to confine the discussion to functions of one variable.

f.2 Probability Densities

It follows from (iii) that if *F(x)* is differentiable at the point *x*, then the probability *density* at the point *x* is defined and is equal to $f(x) = \dfrac{dF(x)}{dx}$; in other words, if you divide the probability that *X* is in a given interval *(x,x + h)* by the length *h* of that interval, and let *h* tend to 0, then if *F* is differentiable there is a probability density at the point *x*, which is equal to *f(x)*. If the density exists at every point, then the associated probability distribution of the random variable is said to be continuous.

Probability densities are of great importance in mathematical statistics—indeed, for many years the principal subject of research in that field was finding the forms of density functions of random variables obtained by transformations of other random variables. They are so important because many of the probability distributions in physics, demography, biology, and similar fields are continuous, or at any rate approximate continous distributions. Few people believe, however, in the real—as opposed to the mathematical—existence of continuous distributions, regarding them as only idealisations of what in fact are discrete distributions.

Many of the famous distribution functions in statistics are identifiable only by means of their associated density functions; more precisely, those cumulative distribution functions have no representation other than as integrals of their associated density functions. Thus the famous Normal Distributions (there is a whole family of these determined by the values of their two parameters, mean *m* and standard deviation σ, which we shall discuss shortly) have distribution functions characterised as the integrals of the density functions.

Another word about notation. For the sake of simplicity,

we shall in the subsequent chapters often use *'P(x)'*, instead of *F(x)* and *f(x)*, to refer to both distribution and density functions. We trust that no confusion will be caused thereby.

f.3 The Mean and Standard Deviation

Two quantities which crop up all the time in statistics are the mean and standard deviation of a random variable X. These are so-called *expected values* of certain functions of X. The expected value of a function $h(X)$ of X is defined to be (where it exists) the probability-weighted average of the values of h. To take a simple example, suppose that h takes only finitely many values $h_1, \ldots, h_n$ with probabilities $f_1, \ldots, f_n$. Then the expected value $E(h)$ of h always exists and is equal to $\sum_{i=1}^{n} h_i f_i$. If X has a probability density function $f(x)$ and h is integrable, then $E(h) = \int_{-\infty}^{\infty} h(x)f(x)dx$, where the integral exists.

The *mean value* of X is the expected value of the random variable X itself; it follows that the mean of X is simply the probability-weighted average of the values of X. The *variance* of X is the expected value of the function $(X - m)^2$, where m is the mean of X. The *standard deviation* of X is the square root of the variance. The square root is taken because the standard deviation is intended as a characteristic measure of the spread of X away from the mean, and so should be expressed in units of X. Thus, if we write s.d. (X) for the standard deviation of X, s.d.$(X) = \sqrt{E((X - m)^2)}$, where the expectation exists.

We can illustrate all the notions in the last few paragraphs with the example of the normal distribution whose mean is m and standard deviation σ. The normal distribution function is given by the integral over the values of the real variable t from $-\infty$ to x of the density we mentioned above, that is, by

$$F(x) = \int_{-\infty}^{x} \frac{1}{\sigma\sqrt{2\pi}} e^{-\frac{1}{2}\left(\frac{t-m}{\sigma}\right)^2} dt$$

It is easily verified from the analytic expression for $F(x)$ that the parameters m and σ are indeed the mean and standard deviation of X. The curve of the normal density is the familiar bell-shaped symmetrical curve about $x = m$ with the points $x = m \pm \sigma$ corresponding to the points of maximum slope of the curve (figure 1). A fact we shall draw on later is that the interval on the x-axis determined by the distance of 1.96 standard deviations centred on the mean supports 95% of the area under the curve, and hence receives 95% of the total probability.

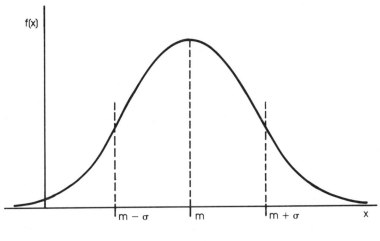

FIGURE 1

■ g PROBABILISTIC INDEPENDENCE

Two sentences h_1 and h_2 are said to be probabilistically independent (relative to some given probability measure P) if and only if $P(h_1 \,\&\, h_2) = P(h_1)P(h_2)$. It follows immediately that, where $P(h_1)$ and $P(h_2)$ are both greater than zero, so that the conditional probabilities are defined, $P(h_1 \mid h_2) = P(h_1)$ and $P(h_2 \mid h_1) = P(h_2)$, just in case h_1 and h_2 are probabilistically independent.

Let us consider a simple example, which is also instructive in that it displays an interesting relationship between probabilistic independence and the so-called Classical Definition of probability. A repeatable experiment is determined by the conditions that a given coin is to be tossed twice and the resulting uppermost faces are to be noted in the sequence in which they occur. Suppose each of the four possible types of outcome—two heads, two tails, a head at the first throw and a tail at the second, a tail at the first throw and a head at the second—has the same probability, which of course must be one quarter. A convenient way of describing these outcomes is in terms of the values taken by two random variables X_1 and X_2, where X_1 is equal to 1 if the first toss yields a head and 0 if it is a tail, and X_2 is equal to 1 if the second toss yields a head and 0 if a tail. According to the Classical Definition, or as we shall call it, the Classical Theory of Probability, which we look at in the next chapter (and which should not be confused with the Classical Theory of Statistical Inference, which we shall also discuss), the probability of the sentence '$X_1 = 1$' is equal to the ratio of the number of those possible outcomes of the experiment which satisfy that sentence, divided by the total number, namely four, of possible outcomes. Thus, the probability of the sentence '$X_1 = 1$' is equal to $\frac{1}{2}$, as is also, it is easy to check, the probability of each of the four sentences of the form '$X_i = x_i$', $i = 1$ or 2, $x_i = 0$ or 1. By the same Classical criterion, the probability of each of the four sentences '$X_1 = x_1 \,\&\, X_2 = x_2$' is $\frac{1}{4}$. Hence

$$P(X_1 = x_1 \,\&\, X_2 = x_2) = P(X_1 = x_1)P(X_2 = x_2)$$

and consequently the pairs of sentences '$X_1 = x_1$', '$X_2 = x_2$' are probabilistically independent (we have avoided answering, or trying to answer here, the question of what criteria justify the application of the Classical 'definition'; in the next chapter we shall discuss that definition in more detail).

The notion of probabilistic independence is generalised to n sentences as follows: $h_1, \ldots, h_n$ are said to be probabilistically independent if and only if for every subset $h_{i_1}, \ldots, h_{i_k}$ of $h_1, \ldots, h_n$,

$$P(h_{i_1} \,\&\, \ldots \,\&\, h_{i_k}) = P(h_{i_1}) \times \ldots \times P(h_{i_k}).$$

It is easy to see, just as in the case of the pairs, that if any set of sentences is probabilistically independent then the probability of any one of them conditional on any of the others, where the conditional probabilities are defined, is the same as its unconditional probability. It is also not difficult to show (and it is, as we shall see shortly, important in the derivation of the binomial distribution) that if $h_1, \ldots, h_n$ are independent, then so are all the 2^n sets $\pm h_1, \ldots, \pm h_n$, where $+h$ is h and $-h$ is $\sim h$.

Any n random variables $X_1, \ldots, X_n$ are said to be independent if for all sets of intervals $I_1, \ldots, I_n$ of values of $X_1, \ldots, X_n$ respectively, the sentences $X_1 \,\epsilon\, I_1, \ldots, X_n \,\epsilon\, I_n$ are probabilistically independent. We have, in effect, already seen that the two random variables X_1 and X_2 in the example above are probabilistically independent. If we generalise that example to that of the coin's being tossed n times, and define the random variables $X_1, \ldots, X_n$ just as we defined X_1 and X_2, then again a consequence of applying the Classical 'definition' to this case is that $X_1, \ldots, X_n$ are probabilistically independent. It is also not difficult to show that a necessary and sufficient condition for any n random variables $X_1, \ldots, X_n$ to be independent is that

$$F(x_1, \ldots, x_n) = F(x_1) \times \ldots \times F(x_n)$$

where $F(x_1, \ldots, x_n)$ is the joint distribution function of the variables $x_1, \ldots, x_n$ (for a proof of this see Cramér, 1946, pp. 159–160).

■ h CONDITIONAL DISTRIBUTIONS

Versions of (17) and (18), the two forms of Bayes's Theorem, can be proved for probability densities. First we note that according to the conditional probability axiom, axiom 4,

(20) $P(X < x \mid y \leq Y < y + \delta y) =$

$$\frac{P(X < x \ \& \ y \leq Y < y + \delta y)}{P(y \leq Y < y + \delta y)}$$

The left-hand side is an ordinary conditional probability. Note that if $F(x)$ has a density $f(x)$ at the point x, then $P(X = x) = 0$ at that point. We noted in the discussion of (4) that $P(a \mid b)$ is in general only defined if $P(b) > 0$. However, it is in certain cases possible for b to be such that $P(b) = 0$ and for $P(a \mid b)$ to take some definite value. Such cases are afforded where b is a sentence of the form $Y = y$ and there is a probability density $f(y)$ at that point. For then, if the joint density $f(x,y)$ also exists, then multiplying top and bottom in (20) by δy, we can see that as δy tends to 0, the right-hand side of that equation tends to the quantity

$$\frac{\int_{-\infty}^{x} f(u,y)du}{f(y)}$$

which determines a distribution function for X, called the conditional distribution function of X with respect to the event $Y = y$. Thus in such cases there is a perfectly well-defined conditional probability

$$P(x_1 \leq X < x_2 \mid Y = y),$$

even though $P(Y = y) = 0$.

The quantity $\dfrac{f(x,y)}{f(y)}$ is the density function at $X = x$ of this conditional distribution (the point $Y = y$ being regarded now as a parameter), and is accordingly called the conditional probability density of X at x, relative to the event $Y = y$. It is of great importance in mathematical statistics and it is customarily denoted by the symbol $f(x \mid y)$. Analogies of (17) and (18), the two forms of Bayes's Theorem, are now easily obtained for densities: where the appropriate densities exist

$$f(x \mid y) = \frac{f(y \mid x)f(x)}{f(y)}$$

$$\text{and } f(x \mid y) = \frac{f(y \mid x)f(x)}{\int_{-\infty}^{\infty} f(y \mid x)f(x)dx}.$$

■ I THE BINOMIAL DISTRIBUTION

Let us end this brief outline of that part of the mathematical theory of probability which we shall have occasion to use, with the derivation and some brief discussion of the limiting properties of the first non-trivial random variable distribution to be investigated thoroughly, the binomial distribution. It was through examining the properties of this distribution that the first great steps on the road to modern mathematical statistics were taken, by James Bernoulli who proved (in *Ars Conjectandi,* published posthumously in 1713) the first of the limit theorems for sequences of independent random variables, the so-called *Weak Law of Large Numbers,* and Abraham de Moivre, an eighteenth century Huguenot mathematician settled in England, who proved that, in a sense we shall make clear shortly, the binomial distribution tends for large n to the normal. Although Bernoulli demonstrated his result algebraically, it follows, as we shall see, from de Moivre's limit theorem.

Suppose $X_i, i = 1, \ldots, n$, are random variables which take two values only, which we shall label 0 and 1, and that the probability that each takes the value 1 is the same for all i, and equals p:

$$P(X_i = 1) = P(X_j = 1) = p.$$

Suppose also that the X_i are independent:

$$P(X_1 = x_1 \& \ldots \& X_n = x_n) =$$
$$P(X_1 = x_1) \times \ldots \times P(X_n = x_n),$$

where $x_i = 1$ or 0. In other words, the X_i are independent, identically distributed random variables. Let $Y = X_1 + \ldots + X_n$. Then for any r, $0 \le r \le n$,

(21) $P(Y = r) = {}^nC_r p^r (1 - p)^{n-r}$

since using the additivity property, the value of P is obtained by summing the probabilities of all conjunctions

$$X_1 = x_1 \& \ldots \& X_n = x_n$$

where r of the x_i are ones and the remainder are zeros. There are ${}^nC_r$ of these, where ${}^nC_r$ is the number of ways of selecting r objects out of n, and is equal to $\dfrac{n!}{(n - r)!r!}$. The term $n!$ is equal to $n \times (n - 1) \times (n - 2) \times \ldots \times 2 \times 1$, and $0!$ is set equal to 1. By the independence and constant probability assumptions, the probability of each conjunct in the sum is $p^r(1 - p)^{n-r}$, since $P(X_i = 0) = 1 - p$.

If Y is any sum of n binomial random variables and satisfies (21), then Y is said to possess the binomial distribution. The mean of Y is np, as can be easily seen from the facts that

$$E(X_1 + \ldots + X_n) = E(X_1) + \ldots + E(X_n)$$

and that

$$E(X_i) = p \times 1 + (1 - p) \times 0 = p.$$

The squared standard deviation, or variance of Y, is

$$E(Y - np)^2 = E(Y^2) + E(np)^2 - E(2Ynp)$$
$$= E(Y^2) + (np)^2 - 2npE(Y)$$
$$= E(Y^2) - (np)^2.$$

Now $E(Y^2)$

$$= \sum E(X_i^2) + \sum_{i \ne j} E(X_i X_j)$$

$$= np + n(n - 1)p^2$$

Hence s.d. $(Y) = \sqrt{-np^2 + np} = \sqrt{np(1 - p)}$.

The far-reaching significance of these expressions is apparent when n becomes very large. De Moivre showed that for large n, Y is approximately normally distributed with mean np and standard deviation $\sqrt{np(1-p)}$ (the approximation is very close for quite moderate values of n). This implies that the variable $\dfrac{(Y-np)}{\sqrt{np\,(1-p)}}$ is approximately normally distributed, for large n, with mean 0 and standard deviation 1. Hence

$$P(-k < Z < k) \approx \Phi(k) - \Phi(-k)$$

where Φ is the normal distribution function with zero mean and unit standard deviation. Hence

$$P\!\left(p - k\sqrt{\frac{pq}{n}} < \frac{Y}{n} < p + k\sqrt{\frac{pq}{n}}\right) \approx \Phi(k) - \Phi(-k),$$

where $q = 1 - p$. So, setting $\epsilon = k\sqrt{\dfrac{pq}{n}}$,

$$P\!\left(p - \epsilon < \frac{Y}{n} < p + \epsilon\right) \approx \Phi\!\left(\epsilon\sqrt{\frac{n}{pq}}\right) - \Phi\!\left(-\epsilon\sqrt{\frac{n}{pq}}\right).$$

Clearly, the right-hand side of this equation tends to 1, and we have obtained the Weak Law of Large Numbers:

$$P\!\left(\left|\frac{Y}{n} - p\right| < \epsilon\right) \to 1, \text{ for all } \epsilon > 0.$$

This is one of the most famous theorems in the history of mathematics. James Bernoulli proved it originally by purely combinatorial methods. It took him twenty years to prove, and he called it his "golden theorem". It is the first great result of the discipline now known as mathematical statistics and the forerunner of a host of other limit theorems of probability. Its extramathematical significance lies in the fact that sequences of independent binomial random variables with constant probability, or Bernoulli sequences as they are called, are thought to model many types of sequence of repeated stochastic trials (the most familiar being tossing a coin n times and registering the sequence of heads and tails produced). What the theorem says is that for such sequences of trials the relative frequency of the particular character concerned, like heads in the example we have just mentioned, is with arbitrarily great probability going to be situated arbitrarily close to the parameter p.

The Weak Law, as stated above, is only one way of appreciating the significance of what happens as n increases. As we saw, it was obtained from the approximation

$$P(p - k\sqrt{\frac{pq}{n}} < \frac{Y}{n} < p + k\sqrt{\frac{pq}{n}}) \approx \Phi(k) - \Phi(-k),$$

where $q = 1 - p$, by replacing the variable bounds (depending on n) $\pm k\sqrt{\frac{pq}{n}}$ by ϵ, and replacing k on the right-hand side by $\epsilon\sqrt{\frac{n}{pq}}$. The resulting equation is equivalent to the first. In other

words, the Weak Law can be seen either as the statement that if we select some fixed interval of length 2ϵ centred on p, then in the limit as n increases, all the distribution will lie within that interval, or as the statement that if we first select any value between 0 and 1 and consider the interval centered on p which carries that value of the probability, then the endpoints of the interval move towards p as n increases, and in the limit coincide with p.

Throughout the eighteenth and nineteenth centuries people took these facts to justify inferring, from the observed relative frequency of some given character in long sequences of apparently *causally* independent trials, the approximate value of the postulated binomial probability. While such a practice may seem *suggested* by Bernoulli's theorem, it is not clear that it is in any way justified. While doubts were regularly voiced over the validity of this 'inversion', as it was called, of the theorem, the temptation to see in it a licence to infer to the value of p from 'large' samples persists, as we shall see in Chapter 9, where we shall discuss the issue in more detail.

The Weak Law of Large Numbers is only one of the theorems of mathematical probability theory which go under the name of 'laws' of large numbers. Another, equally famous for pointing to a connection between probabilities and frequencies in sequences of identically distributed, independent binomial random variables, is the Strong Law, which is usually stated as a result about actually infinite sequences of such variables: it asserts that with probability equal to 1, the value of the sum to infinity Y of the X_i exists (that is to say, the relative frequency of ones converges to some finite value) and is equal to p. So stated, the Strong Law requires for its proof the axiom of countable additivity, though there is a version for finite sequences of random variables which needs only finite additivity (*see,* for example, Feller, 1950, p. 203). We shall leave the discussion of the mathematics of probability there, and now turn our attention to the foundations of the Bayesian methodology and to a discussion of an epistemic interpretation of probabilities.

CHAPTER 3

Subjective Probability

■ a INTRODUCTION

The purpose of this book is to explain characteristic features of scientific inference in terms of the probabilities, relative to the available data, of the various explanatory hypotheses under consideration. In this chapter we shall consider the nature of these probabilities. We shall first discuss, and reject, the thesis, associated in this century with the names of Keynes and Carnap, that they are logical quantities, determined in a quite objective way by the logical structure of the hypotheses and the data. We shall argue that this idea is untenable and that they should instead be understood as subjective assessments of credibility, regulated by the requirement that they be overall consistent; and we will show that a necessary condition for consistency is agreement with the rules of the probability calculus.

The first developed theory of probability was what we now know as the Classical Theory, but which is, as we shall show presently, better regarded as the forerunner of Keynes's and Carnap's explicitly logical theories. The Classical Theory sought to provide a foundation both for the gambler's calculations of his expected gains and losses, and for the philosopher's and scientist's belief in the validity of inductive inference; let us see how.

■ b THE CLASSICAL THEORY

The theory was first expounded in the late seventeenth century (it appears, for example, in James Bernoulli's *Ars Conjectandi*, 1713). But the account which became classic, and one of the most widely read of all works ever written on probability, is Laplace's *Philosophical Essay on Probabilities*, 1820, published originally as the preface to his monumental *Analytical Theory of Probabilities*. The basic principle of the theory was that for suitable factual hypotheses *h* and data *e*, there is a measurable, numerical degree of certainty ("degree of certainty" is actually Bernoulli's term) which should be entertained in the truth of *h* in the light of *e*. These degrees of certainty are located in the closed unit interval, where they are given uniquely as the values of a probability function, with 1 corresponding to the certainty that *h* is true, and 0 to the certainty that *h* is false. Where the probability in any given case could be calculated, its value, call it *p*, was also regarded as providing, in the ratio

$\dfrac{p}{(1-p)}$, the fair odds on *h* in the light of *e*, or the odds relative

to which the rational expectation of gain for each side of a bet at those odds is zero (we shall discuss this notion of fair odds later in this chapter).

The pioneers did not use modern notation, nor did they distinguish explicitly between conditional and unconditional probabilities; only in the work of Keynes and later authors, like Jeffreys (1961) and more recently Carnap (1950, 1952, and 1971), do we find such probabilities represented explicitly as conditional probabilities. Nevertheless, it seems clear that after the discovery of Bayes's Theorem and the use to which it was put by Laplace, the probability function which he himself took to measure the rational degree of certainty of *h* in the light of *e* was a conditional probability. We shall in what follows represent these probabilities by the modern notation *P(h | e)*.

b.1 The Principle of Indifference

Implicit in the Classical Theory is the principle that if there are *n* mutually exclusive possibilities $h_1, \ldots, h_n$, and *e* gives no more reason to believe any one of these more likely to be true than any other, then *P(h_i | e)* is the same for all *i*. The principle was later called by von Kries (1886) the Principle of Insufficient Reason, and by Keynes (1921) the Principle of Indifference (we shall follow Keynes's terminology, merely because it is more compact). It is of fundamental importance within the Classical Theory, because the judgment that *e* is indifferent, as it were, or epistemically neutral, between a set of exclusive alternatives seems to be one which can be made easily in certain cases. If, for example, *e* were to say merely that a number between 1 and *n* will be selected, and nothing else, then it seems plausible at the least that e is indifferent between the hypotheses of the form 'i is selected', for each integer i between 1 and n. Let us see what follows, at any rate, from supposing that there are sets of alternatives $h_1, \ldots, h_n$ between which such judgments of equality are valid. In particular, let us suppose that the h_i are also an exhaustive set and that h is a hypothesis equivalent, given e, to a disjunction of r of them. In Laplace's now famous terminology, the h_i represent "cases equally possible" in the light of e, and the r hypotheses whose disjunction is equivalent to h are the 'cases favourable to h'. It follows from (14)–(16), Chapter 2, that $P(h_i \mid e) = \dfrac{1}{n}$, for all i, and that $P(h \mid e) = \dfrac{r}{n}$. Let us quote Laplace:

> The theory of chance consists in reducing all the events of the same kind to a certain number of cases equally possible, that is to say, to such as we may be equally undecided about in regard to their existence, and in determining the number of cases favourable to the event whose probability is sought. The ratio of this number to that of all the cases possible is the measure of this probability, which is thus simply a fraction whose numerator is the number of favourable cases and whose denominator is the number of all the cases possible. (Laplace, 1820, pp. 6–7).

The earliest applications of the Classical Theory were to games of chance, like rolling dice or picking tickets from an urn or tossing coins, because their classes of elementary outcomes are finite and also, as a consequence of the carefully randomized structure of those games, we seem very much to be in a prior state of "equal indecision", to use Laplace's terminology, as to the occurrence of each of those outcomes. Later, in the pioneering work of Bayes (1763), the scope of the Classical Theory was extended to outcome spaces which could be represented by a bounded one- or many-dimensional interval of real numbers: if all the points in the interval $[a,b]$ are the possible values of a real parameter T, deemed equiprobable in the light of our background data, then the probability that T will lie in a subinterval $[c,d]$ is plausibly $\dfrac{|d - c|}{|b - a|}$.

It might be objected that the randomizing procedures used in games of chance, like shuffling decks of cards or making tossed coins as nearly perfectly balanced as physically possible, are

regarded as legitimate grounds for distributing equal degrees
of certainty over the elementary outcomes of the apparatus
involved only after we have observed that such procedures are
as a matter of fact associated with roughly equal frequencies
in those outcomes. We are therefore, so the objection continues,
appealing not to a criterion of 'epistemic neutrality' at all,
but covertly to an unjustified equation of degree of certainty
with long-run observed frequency.

The Classical theorists after Laplace had an answer to this,
which was that the objection puts the cart before the horse.
Laplace, building on earlier work of Bayes, seemed in a famous
demonstration to have shown that the identity between observed frequencies and degrees of certainty is not an assumption, covert or otherwise, which needs to be made, but on the
contrary appears to be a consequence of assuming that at the
outset one knows *nothing whatever* about the propensity of the
apparatus concerned for generating different frequencies of outcomes. If one assumes literally nothing about such propensities,
then it would seem a legitimate application of the Principle of
Indifference that the various frequencies are of equal epistemic
standing relative to this null data set. It was on the basis of
this application of the Principle of Indifference that Laplace
was able to prove his result, the famous, or infamous, Rule of
Succession. In so doing he seemed also to have answered successfully Hume's famous thesis that one cannot without cir-
cularity prove that the observation of past events gives any
guidance to what will occur in the future. Because Laplace's
result appears to provide a justification for a very strong type
of enumerative induction, one moreover which in this century
has been accorded the status of a purely logical principle, and
because it introduces in an informal way the central ideas of
Carnap's later, explicitly logical theory, we shall give an outline
of Laplace's argument.

b.2 The Rule of Succession

Laplace's result was called the Rule of Succession by John Venn
(1866), and the name stuck. It states that if an event A is one
of the possible outcomes of instantiating some set of repeatable
conditions, and if in m repetitions A is observed to occur r times,
then the probability that it will occur at the $(m + 1)$th repetition, conditional upon the data, is $\dfrac{(r + 1)}{(m + 2)}$. Since as m grows

large $\dfrac{(r + 1)}{(m + 2)}$ is approximately equal to $\dfrac{r}{m}$, the Rule of Succession appears to provide a justification for estimating epistemic probabilities by sample frequencies; hence, when $r = m$, that is to say when we have observed only A's in the sample, $\dfrac{(r + 1)}{(m + 2)}$ tends quickly to 1, the value of certainty.

Laplace obtained the result by considering a random sample, without replacement, from an 'infinite urn' containing black and white tickets in an unknown 'proportion' p. A proportion in an infinite set is obviously a fiction unless understood in some limit sense, and Laplace's result is in fact a mathematically, if not philosophically, quite respectable limiting result obtained by letting the number of tickets in a finite urn grow very large. We shall reconstruct his reasoning in terms of a large finite urn. The reason for considering very large numbers of tickets at all is that Laplace wanted, as we shall see, to use urn-sampling as a model of an experiment with outcomes A and not-A, which is capable of being repeated an arbitrary number of times.

Suppose the urn contains n tickets, q of which are white; m tickets are withdrawn, r of which are white. What is the probability, relative to this information, that the next ticket to be withdrawn will be white? Applying the Principle of Indifference first to the probability of drawing r white tickets out of m sampled, without replacing any of them during the sampling, conditional upon the data that r of the tickets are white, we obtain the value

$$\frac{{}^{m}C_r \, {}^{n-m}C_{q-r}}{{}^{n}C_q}.$$

Let $p = \dfrac{q}{n}$. Then for large n this expression is approximately equal to ${}^{m}C_r p^r (1 - p)^{m-r}$. The proportion p of white tickets will take one of the values $0, \dfrac{1}{n}, \dfrac{2}{n}, \ldots, 1$, and we next apply the Principle of Indifference to these $n + 1$ possibilities, giving the (unconditional) probability of each possibility as $\dfrac{1}{(n + 1)}$. By the Theorem of Total Probabilities proved in Chapter 2, the (unconditional or prior) probability of drawing r white tickets out of a sample of size m is obtained by multiplying these two probabilities: the probability of obtaining r white tickets out of m drawn conditional on there being altogether q white tickets in the urn, and the probability of there being q white tickets in the urn, and summing the result over all values of q from 0 to n. Since the prior probability distribution $\dfrac{1}{(n+1)}$ generates a uniform limiting probability density equal to 1, the total probability of getting r white tickets out of m is approximately

$$\int_0^1 {}^mC_r p^r (1-p)^{m-r} dp.$$

The probability of the $(m+1)$th ticket being white is, by analogous reasoning, approximately equal to

$$\int_0^1 {}^mC_r p^{r+1}(1-p)^{m-r} dp,$$

and so, by axiom 4 of the probability calculus (Chapter 2, section **c.1**), the probability, given that one has drawn r white tickets out of m, of the next ticket to be drawn being white is equal, in the limit as n tends to infinity, to

$$\frac{\displaystyle\int_0^1 p^{r+1}(1-p)^{m-r} dp}{\displaystyle\int_0^1 p^r(1-p)^{m-r} dp}.$$

This quantity is equal to $\dfrac{(r+1)}{(m+2)}$. This is the Rule of Succession. Despite the apparently asymptotic nature of the result, it is actually quite independent of n, as Carnap's work was later to show (though the independence of n was, according to Jeffreys, 1961, p. 127, first noticed by C. D. Broad).

Laplace regarded a selection from his very large urn as a model of making an observation to see whether A occurs or not. An immediate corollary of the Rule of Succession is that if that event has occurred at every observation $(r = m)$, then the probability that it will occur at the next tends quite quickly to 1. Using this corollary, Laplace, in a much-quoted passage, proceeded to compute the exact odds on the sun's rising the next day ("It is a bet of 182614 to one that it will rise again tomorrow." Laplace, 1820, p. 19). Although modelling arbitrary experiments as random selections without replacement from infinite urns might be thought highly questionable, when we come to consider Carnap's work, we see that the same formulas apply without employing this or any other concrete model. At any rate, the use of urn models as analogues of actual physical experiments became standard practice in the nineteenth century.

b.3 The Principle of Indifference and the Paradoxes

Despite an almost universal following in the last century, by the end of the first half of this one the Classical Theory had very few influential advocates. This seems to have been due to two causes. One was the cavalier use that was made of the Rule of Succession and a growing recognition that many of its results were intuitively absurd (for example, tossing a coin once and observing a head makes the probability of heads at the next toss $\frac{2}{3}$). Keynes, in a memorable passage, wrote

> No other formula in the alchemy of logic has exerted more astonishing powers. For it has established the existence of God from total ignorance, and it has measured with numerical precision the probability that the sun will rise tomorrow. (Keynes, 1921, p. 89)

The second, probably more important, reason for the discrediting of the Classical Theory was that the Principle of Indifference on which it was based was discovered to yield inconsistencies with alarming ease. For example, suppose that you draw a ball from an urn which you are told contains white and coloured balls in some unknown proportion, and that the coloured balls are either red or blue. What sort of ball will you draw? Your data seem to be neutral, in the first instance, between its being white or coloured. Hence according to the Principle of Indifference, the probability that it will be white is one half. But if it is coloured, the ball is red or blue, and the data are surely neutral between the ball's being either white or blue or red. Hence according to the Principle of Indifference, the probability of the ball's being white is one third.

'Paradoxes' (as they came to be called) of this type abound. Consider another, which occasioned a good deal of discussion. In the example of Laplace's urn it seemed a straightforward application of the Principle of Indifference to ascribe equal probabilities to the possible proportions of white tickets. But is it?

After all, if we distinguish the individual tickets by numbering them from 1 to n, then the Principle of Indifference appears to imply that the probabilities of the various possible proportions are *not* all equal. For suppose q of the tickets are white. There are ${}^nC_q$ different ways in which the property 'being white' can be distributed among q of the tickets. There are, therefore, $\sum_{q=0}^{n} {}^nC_q = 2^n$ of these distinct possible constitutions of the urn, none of which do our data tell us is any more likely than another. Consequently each constitution should receive the probability 2^{-n}. But that means that the probability of m of the tickets being white is ${}^nC_m 2^{-n}$, which varies with m; and so the probabilities of the various possible proportions are not equal. (Actually, Laplace *had* assumed that the tickets were individually distinguishable, so his derivation of the Rule of Succession is founded upon an inconsistent use of the Principle.)

Bertrand's well-known paradoxes of geometrical probability showed that the Principle of Indifference generated inconsistencies also in the continuous domain (a very clear presentation is given by Neyman, 1952, pp. 15–17). A simple example of the type of inconsistency which can be generated by continuous variables is afforded by considering a parameter T about which nothing is known except that its value lies in some specified bounded interval $[a,b]$, $0 < a$. Suppose that we take our essentially null information to justify a prior density for T equal to $(b - a)^{-1}$. However, if all we know about T are the limits of its possible values, then all we know about the parameter T^2 are the limits of its possible values. T^2 should therefore possess, by the same token, a uniform prior density equal to $(b^2 - a^2)^{-1}$. However, we now have a contradiction; for it follows from the assumption that the density of T is uniform in $[a,b]$ that the density of T^2 is not uniform, but is equal to $\dfrac{2x}{(b - a)}$ at the point $T = x$.

The idea of a probabilistic logic of inductive inference based on some form of the Principle of Indifference nevertheless retained a powerful appeal. Keynes (1921) continued in the programme inaugurated by Leibniz and Bernoulli, but elaborated and sharpened their thesis that the probability calculus is a more general logic than the logic of deduction with the claim that between any two propositions a and b there is a relation of partial entailment, such that the degree to which a entails b is measured by a unique conditional probability $P(b \mid a)$, and this degree of *partial entailment* is equal to the degree of belief which it would be rational to repose in b were your data restricted to a (Keynes, 1921, p. 16; Keynes also thought, however, that $P(b \mid a)$ did not always take numerical values, and indeed that probabilities in general are only partially ordered).

When it came to evaluating these ostensibly logical probabilities for those pairs of statements for which P did take numerical values, Keynes recommended a form of the Principle of Indifference which he hoped retained what he regarded as its valid core, without however generating paradoxes and inconsistencies. His modified version of the principle is as follows. Suppose (a) that the data e determine a finite set of n exclusive possibilities h_i, (b) that e has 'the same form' relative to each of the h_i, and (c) that none of those hypotheses is further 'divisible' or decomposable into subalternatives of the same form as some other hypothesis in the original set. Then and only then should the h_i be assigned equal probabilities relative to e (Keynes, 1921, pp. 60–66). This modification seems to meet the difficulty generated by cases such as the urn containing white and coloured balls. For it is now no longer legitimate to infer that the prior probabilities of the drawn ball's being either white or coloured are equal, since the latter possibility decomposes into a disjunction of two narrower possibilities, of the same simple colour kind as the former, being red or being blue.

But it is not clear that this modification of the Principle of Indifference saves it from its problems with continuous transformations of random variables whose possible values are an interval on the line. Keynes asserted that a legitimate application is to divide up the interval (if it is bounded) into a finite number m of subintervals of equal length, and to assign equal probabilities to each subinterval, letting m then tend to infinity. But he conceded that the resulting uniform density distribution is not maintained under a large class of transformations, a fact which he attempted to dismiss by saying that it is an example of the phenomenon, well known to mathematical physicists, where often the form of a density distribution depends on the passage to the limit (1921, p. 69). This is true, but does not save him the arbitrary choice of how in any given case to proceed to the limit (that is, which particular variable to take as determining the uniform distribution).

Also, it is far from clear that the colour categories in the example above *are* of the same kind: white light, for example, is a superposition of light of other wavelengths. 'Being white' is a colour category of the same form as 'being blue' only relative to a particular linguistic system, which does not make distinctions which might be made by a more refined one. So there seems to be implicit in Keynes's analysis reference to some appropriate language, sufficiently well defined syntactically to permit judgments of sameness or difference of form of its component predicates.

Keynes, writing in England in the second decade of this century, was geographically and temporally just too distant from the Continental work, still in its pioneering stage, which would revolutionise the study of formal logic and give a precise sense to the sorts of syntactic criteria he attempted to enunciate. Thirty years later Carnap was able to remedy the deficiency, and in so doing find reasons, which seemed to him conclusive, against adopting the Keynesian form of the Principle of Indifference. What Carnap was not able to do, however, was eliminate the arbitrariness inherent in the language-relativity of the probability distributions he found more congenial.

■ c CARNAP'S LOGICAL PROBABILITY MEASURES

c.1 Carnap's c† and c*

Carnap, in his first detailed (and monumental) work on probabilistic inductive logic (1950), attempted to exploit the advances in logical notation and theory made in the previous decades to create a purely logical theory of induction, in which the value of an appropriate conditional probability function, which he wrote *c(h,e)*, measures the degree of rational credibility of *h* relative to data or evidence *e*. While in the work cited he deliberately avoided the term 'degree of rational belief', because of what he called its "psychologism", and regarded *c(h,e)* as a formal explication of, among other things, the degree of confirmation of *h* by *e*, there is no doubt that his work is a continuation of the Bernoulli-Laplace-Keynes tradition. For *c(h,e)* is also regarded by Carnap as determining, in the quantity $\dfrac{c(h,e)}{(1 - c(h,e))}$, the fair odds on *h* in the light of *e*; and indeed, the very term 'degree of rational belief' makes an appearance in his later work (1971, p. 9, for example).

Carnap followed Keynes in regarding $c(h,e)$ also as a measure of the extent to which h is entailed by e (though as a matter of fact the idea goes much farther back, originating in Bolzano, 1837). One of the fruits of the development of formal logic between the date of Keynes's *Treatise* and Carnap's *Logical Foundations* was that Carnap was able to demonstrate that any conditional probability function $P(b \mid a)$ defined on the sentences of some language can be understood as determining a measure of the extent to which a entails b. It is easy to see this. According to the standard definition of logical consequence, a is a consequence of b if all the possible structures interpreting a predicate language L (or L-possible worlds) which make b true also make a true. Using the notation of Chapter 2, section **c.3**, let $M(a)$ be the set of all such structures making a true. As we saw there, every probability P on the sentences a of L determines a real-valued, nonnegative, additive function P' on the sets $M(a)$. Such a function is technically a normalised measure. Now any finite measure on the same field obeys the same formal laws as the ordinary counting measure which, if a set A is finite, simply tells us how many things there are in A. Now

$$P(b \mid a) = \frac{P(b \text{ \& } a)}{P(b)} = \frac{kP'(M(a) \cap M(b))}{kP'(M(b))}$$

for all nonzero k, and all b such that $P(b) > 0$. But kP' is a finite measure on the sets $M(a)$, and so the ratio on the right-hand side, and therefore $P(b \mid a)$, can be regarded as defining a generalised 'proportion' of those structures making b true which also make a true, and hence determining a measure of the 'degree' to which b entails a.

But this is not very interesting, since *any* conditional probability function P has this property, independently of what its actual values are. What Carnap attempted to do was to impose constraints of a methodological/epistemological kind on $c(h,e)$ which would eventually determine the function uniquely as the foundation for a quantitative system of inductive logic. This he never succeeded in doing, though he did consider two such functions, called by Carnap $c^\dagger$ and c^*, as candidates for that role. We shall now show, for the simplest type of Carnapian language, how Keynes's modified Principle of Indifference leads directly to one of these functions, $c^\dagger$ which fails conspicuously to meet the methodological criteria Carnap regarded as appropriate to an inductive probability measure. This fact may explain why Carnap never explicitly in his 1950 work or thereafter invoked that principle, though $c*$, the other of the two functions, and the one favoured by Carnap, is, as we shall see, none other than Laplace's measure, generating the Rule of Succession.

Consider a very simple predicate language $L(A,n)$ which enables us to discuss the possession or non-possession of an attribute A by any or all of n distinct individuals a_i. We can list all the completely specified possible states of affairs in such a simple universe by means of the conjunctions

$$\pm A(a_1) \ \& \ \pm A(a_2) \ \& \ \ldots \ \& \ \pm A(a_n),$$

where $+A(a_i)$ is $A(a_i)$ and $-A(a_i)$ is $\sim A(a_i)$. There are clearly 2^n such sentences, called the state-descriptions of $L(A,n)$ by Carnap. These state-descriptions are complete specifications of the sorts of structure describable within $L(A,n)$; hence they constitute an ultimate partition relative to the descriptive resources available within $L(A,n)$: they are not further decomposable within $L(A,n)$. We may therefore apply the Keynesian form of the Principle of Indifference to infer that the probabilities of each of the state-descriptions relative to null data are equal to 2^{-n}. Such probabilities are unconditional probabilities (though they can of course always be represented as conditional probabilities relative to a tautology: $P(h)$ is always equal to $P(h \mid t)$, where t is a tautology). Carnap represented the unconditional probability of a sentence h by $m(h)$, calling m the measure function associated with c. The m which assigns the state-descriptions the equal weights of 2^{-n} he wrote as $m^\dagger$, and the conditional probability function $c(.\,,\,.)$ based on $m^\dagger$ is $c^\dagger$.

Any sentence h of $L(A,n)$ can be represented as a disjunction of q of these state-descriptions, where q is a number between 0 and 2^n inclusive. Hence $m^\dagger(h) = \dfrac{q}{2^n}$. If h says that j of the a_i are A's, then it is easy to see that its unconditional probability is $\dfrac{{}^nC_j}{2^n}$. In $L(A,n)$ there is a sentence $e(k,r)$ which states that r of the individuals $a_1, \ldots, a_k$ possess A. Such a sentence can be taken to describe a sequence of k successive observations, r of which exemplify A. Using Carnap's symbolism,

$$c(A(a_{k+1}),e(k,r)) = \frac{m(A(a_{n+1}) \& e(k,r))}{m(e(k,r))}$$

for all conditional probabilities c, and it is not difficult to work out that for c^r this is equal simply to $\frac{1}{2}$, *independent of both k and r*. In other words, no sample parameters convey any information about the likelihood of the next individual to be observed being an A. This can be stated in terms of Laplace's urn as follows: if the constitutions of the urn, meaning all possible distributions of the property of being white among n numbered tickets, are all equally probable, then all the statements of the form 'the ith ticket is white (not white)', $i = 1, \ldots, n$, are probabilistically independent. Laplace's use of the Principle of Indifference to justify the equiprobability of the possible *proportions* of white tickets is now seen to be crucial to generating nontrivial inductive inferences on the basis of sample evidence. But we have shown that if the Principle is to apply to the elementary possibilities relative to a language with names for all the individuals, then such inferences are impossible. Keynes himself was quite clear that the Principle should apply *only* to the state-descriptions: he wrote that "the equiprobability of each 'constitution' [i.e., m^r] is alone legitimate, and the equiprobability of each numerical ratio erroneous" (1921, p. 61).

Carnap regarded c^r as a totally unsuitable foundation for his system of quantitative inductive logic, and hence in effect abandoned the Keynesian form of the Principle of Indifference. He opted rather to follow Laplace (though he did not say this explicitly) in making equiprobable the possible proportions of A's in the universe, and then, for each q between 0 and n inclusive, making equiprobable all the ^nC_q possible ways in which q of the n individuals may be A's. The measure function m which does this Carnap called m^*, and the c function based on it is c^*. The sentences of $L(A,n)$ describing the possible proportions of A's in the universe are equivalent to those disjunctions of all state-descriptions containing the same number of unnegated A's; these disjunctions Carnap called the *structure-descriptions* of $L(A,n)$. (This terminology is misleading since, as we saw, the state-descriptions more truly determine the types of structure describable within $L(A,n)$). Of course, c^* is the Laplacean probability function, and so Carnap was able to derive the Rule of Succession (for all values of n, as we noted earlier) for c^*, and consequently to obtain probability values for predictions, which are positively sensitive to the sample parameters k and r.

c.2 The Dependence on A Priori Assumptions

c^* is simply one among many other conditional probability functions whose values are sensitive to sample data. Carnap was aware of this and conceded that any choice of one among these was bound to be arbitrary to a greater or lesser extent: c^t and c^* were merely the two functions "which are most simple and suggest themselves as the most natural ones". Indeed, all he could find to say positively in favour of c^* was that it "is the only one which is not entirely inadequate" (1950, p. 565). In his 1952 book he considered a continuum of other 'inductive methods', or conditional probability functions defined on simple languages like $L(A,n)$, each method corresponding to a value of a real-valued non-negative parameter λ; c^* corresponds to setting λ equal to 2 for $L(A,n)$, while the measure assigning equal probabilities to all the state-descriptions corresponds to setting λ equal to infinity. Later (Carnap and Jeffrey, 1971), he considered an even more extensive class of "credibility" measures (his term), defined relative to very much more expressive, set-theoretical languages.

Hintikka (1965, 1966, 1968, and elsewhere) and his followers have experimented with more general first-order languages than Carnap's original monadic ones, where now no bound is placed upon the size of the universe of discourse and where consequently the classes of L-elementary possibilities cannot be characterised in terms of state-descriptions. In particular, Hintikka sought ways of assigning a priori probabilities (the unconditional probabilities corresponding to Carnap's measure-functions m) to these possibilities in such a way that universal hypotheses can have positive probabilities; in Carnap's λ-continuum they are all assigned zero probability by all positive values of λ, and Hintikka and many others regarded this as unacceptable in a theory which proposes to justify inductive procedures—for we certainly do regard universal laws as supported by appropriate experimental evidence.

Whereas Carnap's Continuum of Inductive Methods is parametrised just by one real parameter, λ, Hintikka introduced more. The resulting systems are rather complex (though for an excellent simplifying discussion, *see* Kuipers 1978 and 1980), and we shall only sketch their features salient to this discussion. In these systems the elementary possibilities are the so-called *constituents* of the language. A constituent is a sentence of the language that says which predicates, including those definable within the language in terms of the primitive predicates, identity, and any individual names there might be, are instantiated and which not. There are usually infinitely many L-definable predicates, where L is the (usually first-order) language, each of which can be characterised by the minimum degree of complexity of its defining formula. A given degree of complexity also determines, therefore, a particular set of constituents, and a result due to Hintikka himself is that every sentence in a first-order language is equivalent to a disjunction of finitely many constituents of a given complexity. Since the constituents are mutually exclusive, any set of a priori probabilities assigned to the constituents in L then determine the probabilities, and the conditional probabilities, of all other sentences in L.

The advantage of distributing probabilities via constituents rather than state-descriptions is that state-descriptions are easily defined only for simple monadic languages, whereas the former method imposes no such restrictive constraints. Also, it is easy to see, using the constituent method, how to distribute probabilities in such a way as to endow universal sentences with positive a priori probabilities. Consider the language possessing only one predicate symbol, B say, and no individual names or identity. There are three constituents only in this language, namely the statement that all the individuals are B; the statement that some individuals, but not all, are B; and the statement that no individuals are B. One a priori probability-weighting is that which assigns equal probabilities of one-third to each constituent, giving the two law statements in this language a priori probabilities each of one-third independently of the size of the domain within which they are to be interpreted.

Hintikka and his followers have experimented with various types of a priori distribution over constituents, in languages in general much more sophisticated than this one. The details do not concern us, for we are concerned here with a certain basic strategy common to Carnap and Hintikka, and one which is implicit in the entire programme of constructing a probabilistic epistemology on the lines laid down by Bernoulli and Laplace. For any conditional probability distribution over the sentences of a language necessarily involves the assignment of unconditional probabilities to a partition of the space of possibilities representable within L. But what considerations can possibly justify any such a priori distribution?

One possible answer, and the one implicit in the Keynesian form of the Principle of Indifference, is that the distribution of prior probabilities is legitimate if it is neutral, or, as it is sometimes put, 'informationless', with respect to the class of elementary possibilities determinable within that language: in other words, all the finest-characterised possibilities admitted by that language should receive equal a priori probabilities.

This answer will not do, however. In the first place, such a priori assignments necessarily exhibit a more or less strong bias against certain types of world, which we have no right, in advance of all empirical information, to indulge in. This a priori bias can take extreme forms. For example, consider the classification of worlds according to whether they admit the truth of at least one universal law or whether they admit the truth of none. The measure assigning equal probabilities to the state-descriptions of the language $L(A,n)$ is as far as it is possible to be from being neutral between these two categories of world, telling us that in the limit as n increases, it is with probability one the latter sort of world we live in, for there are only two distinct universal hypotheses in $L(A,n)$: the state-description in which every individual possesses A and the state-description in which no individual possesses A. Their disjunction is therefore equivalent to the statement 'some universal law is true'. But each of them has zero probability with respect to that measure, as is easy to see, and hence so does their disjunction. Indeed, however neutral is any proposed a priori measure as between the members of one partition of possibilities, one can always find another with respect to which that measure is as biased as one likes. All attempts to achieve neutrality in this way will in fact be very partial towards and against types of possible world; uniform neutrality is impossible, just as attempts to iron out a wrinkle in a badly made suit of clothes will not remove it but simply send it elsewhere.

Secondly, elementary possibilities are elementary only relative to some language, and language is a human artifact whose ultimate categories stand on a footing of equality only as a result, therefore, of a collective decision that they should do so, a decision which may consequently be revoked. Often the decision to refine or otherwise amend these categories (and even, possibly, abandon them) is taken as a result of the development of a particular scientific theory and the belief that the latter provides the most adequate representation of some empirical domain to date. But the systems of inductive logic we have been discussing are supposed to adjudicate decisions of empirical adequacy.

That these systems themselves may depend on some prior inductive judgment is not a novel suggestion, however. Carnap has himself suggested (1952, p. 55) that the parameter λ may be evaluated empirically, by a process which today we call *calibration* and which consists in comparing the class of predictions assigned $x\%$ probability with the frequency with which those predictions were true. If there is a significant discrepancy between the probability and the truth-frequency (if the inductive method is not calibrated, in other words), then Carnap recommends adjusting the value of λ appropriately.

But this calls into question the fundamental role assigned his systems of inductive logic by Carnap. If their adequacy is itself to be decided empirically, then the validity of whatever criterion we use to assess that adequacy is in need of justification, not something to be accepted uncritically. As if aware of this objection, Carnap subsequently reverted to his earlier position that "in principle it is never necessary to refer to experience in order to judge the rationality of a c-function" (Carnap, 1968, p. 264; see also Carnap 1971, p. 25). But the position this retreat represents is hardly, if we are correct, any more tenable; for it is that same a priorism we have seen cause to reject.

To sum up: any a priori probability distribution, be it an equiprobability distribution over the state-descriptions of some language or other, or some other distribution, is going to be arbitrary. For this reason we do not regard people who try to evaluate the probabilities of hypotheses relative to data as doing exercises in a genuine logic of generalised deduction, for we take logic to be essentially noncommittal on matters of fact. (Popper has repeatedly made the same general point about Carnap, but this did not stop him advancing his own a priori distribution—*see* Chapter 11, section **c**.) Carnap's claim in the preface to his *Logical Foundations of Probability* that "all principles and theorems of inductive logic are analytic ... hence the validity of inductive reasoning is not dependent upon any synthetic presuppositions" is not true—not, at any rate, of the probabilistic inductive logic he went on to expound.

This seems to leave us with the conclusion that there are no valid objective criteria on which to ground assessments of the probabilities of hypotheses, a conclusion which has led many people to reason further that the probability calculus is incapable of providing a foundation for an objective theory of scientific inference. So it seemed to Fisher, in one of his (very influential) critiques of the Principle of Indifference which, he claimed

> leads to apparent mathematical contradictions. In explaining these contradictions away, advocates of inverse probability [this was the traditional name for the use of Bayes's Theorem to generate posterior probabilities] seem forced to regard mathematical probability . . . as measuring merely psycholog- ical tendencies, theorems respecting which are useless for sci- entific purposes. (Fisher, 1947, pp. 6–7)

It is the burden of the rest of this book that Fisher's infer- ence is incorrect and that theorems respecting such psycholog- ical tendencies are so far from being useless for scientific pur- poses that they form the everyday logic—and a genuine logic at that—of scientific inference. This statement might seem to stand in direct contradiction to our previous remarks. It does not. A first step on the way to showing exactly why it does not is to establish what Fisher himself took for granted, that the "psychological tendencies" of which he spoke—and which we shall characterise explicitly as individuals' degrees of belief— can be numerically represented as mathematical probabilities. To this task we now turn.

■ d DEGREES OF BELIEF AND THE PROBABILITY CALCULUS

d.1 Betting Quotients and Degrees of Belief

Our point of departure is the theory of betting odds, and in particular those odds on a hypothesis h which, so far as you can tell, would confer no positive advantage or disadvantage to anyone betting on, rather than against, h at those odds, on the (possibly counterfactual) assumption that the truth-value of h could be unambiguously decided. Such odds, if you can determine them, we shall call your *subjectively fair* odds on h.

Usually you will be able to determine such odds only within some larger or smaller interval; in what follows, however, we shall assume the ideal case where you can state them as an exact number, and we shall discuss later to what extent this idealisation is permissible.

We should stress that the definition of subjectively fair odds does not presuppose that any odds are fair in fact. We shall discuss later the question of whether any odds are actually fair; all that we are assuming now is that people do, rightly or wrongly, *think* that some odds are fair. And that assumption seems borne out in people's dispositions to bet. This is not to say that the odds they bet at are the ones they find fair. Usually this will not be the case, for it seems that most people bet only when they think the odds advantageous to them. But the very fact that there are predictions h such that people have a tendency to regard one range of odds on h as definitely advantageous to the punter, and another range of odds, in the complement of the first, as advantageous to the bookmaker, entails that—to within some interval, admittedly—some odds are regarded as fair.

Let us continue to idealise and suppose that the odds you take to be fair on h is a single ratio. This number will reflect the extent to which you believe it likely that h will turn out to be true, or what we shall call *your degree of belief* in h. We ought to emphasise that believing certain odds fair does not in any way imply that you will actually accept bets at those or even any greater odds. To believe odds to be fair is to make an intellectual judgment, not, except possibly in special conditions which will vary from individual to individual, to possess a disposition to accept particular bets when they are offered. To stake money on the occurrence of an uncertain event is in effect to buy the option of receiving a larger sum if that event does occur and nothing if it does not. The odds you believe fair determine the price you think fair for that option; but having a belief that a price is fair does not commit you to buying the good. You may not be able to afford the price; or you may simply not want the product at all. We are labouring what is after all a very elementary point because much of the literature on subjective probability is committed to a strongly behaviouralist interpretation of belief, according to which strength of belief is manifested in a willingness to bet at all odds up to those you think justified in the light of your information. Kyburg, for example, writes (1983, p. 64) that "The time-honored way of finding out how seriously someone believes what he says he believes is to invite him to put his money where his mouth is". But that you refuse such an invitation does not at all imply that you do not believe what you say you believe; backing up judgments with financial commitment is a luxury not everyone can afford, even if they wanted to.

To continue: odds are ratios, and hence numbers, and it might seem natural therefore to use these as measures of degrees of belief. However, there is a good reason why the odds scale is a very unsuitable one and in practice not used. The reason is that length of interval does not, in this scale, measure the difference between degrees of belief. The odds scale goes from 0 to plus infinity, with 1 as the point of indifference; hence the difference between being cognitively indifferent between h and $\sim h$ and being certain that $\sim h$ is true is 1, whereas the difference between being certain that h is true and being cognitively indifferent between h and $\sim h$ is infinite. The standard solution to the problem is to transform the semi-infinite odds scale $\mathbb{O}$, with ∞ appended, into the closed unit interval by means of the one-to-one mapping $p = \dfrac{\mathbb{O}}{(1 + \mathbb{O})}$. Odds of 1 (even-money odds) go to $\frac{1}{2}$ under this mapping, 0 goes to 0, and ∞ goes to 1, giving the desired symmetry about the point of indifference between h and $\sim h$.

The function $\mathbb{O} = \dfrac{p}{(1 - p)}$ is inverse to $p = \dfrac{\mathbb{O}}{(1 + \mathbb{O})}$. It follows that to say that one's degree of belief in h is p is equivalent to saying that one's subjectively fair odds on h are $\dfrac{p}{(1 - p)}$.

$\dfrac{\mathbb{O}}{(1 + \mathbb{O})}$ is called the *betting quotient* associated with the odds $\mathbb{O}$. In the light of the discussion above we shall henceforward identify your degree of belief in h with the betting quotient associated with your subjectively fair odds on h.

Characterising degrees of belief, as we have done, in terms of attitudes to actual or potential bets is now so common as to be traditional; and most authors have followed Ramsey's lead in making a willingness actually to bet in suitable circumstances the criterion of strength of belief (Ramsey, 1931, p. 79). This leads, as we shall see later, to severe if not intractable problems when the question is posed why these behaviourally elicited quantities should satisfy the probability calculus. We are not assuming that any type of behavioural display accompanies or is a consequence of possessing a degree of belief. This is not to deny that beliefs have behavioral consequences in certain conditions: they may well, but stating what those conditions are with any precision is a task fraught with difficulty, and fortunately it is not at all necessary to our purposes that we have to do so.

d.2 Why Should Degrees of Belief Obey the Probability Calculus?

In this section we shall prove the central result of this chapter, which is that if a set of betting quotients fails to satisfy the probability calculus, then were anybody to bet indifferently on or against the associated hypotheses, at the odds determined by those quotients, he or she could be made to suffer a net loss (or gain) independently of the truth or falsity of those hypotheses. The importance of this result lies in the corollary, that betting quotients which do not satisfy the probability axioms cannot consistently be regarded as determining fair odds. The corollary follows because (i) fair odds have been characterised as those which assign zero advantage to either side of a bet at those odds, where the advantage is regarded as a function of the odds only and not the stakes; (ii) the sum of finitely (or even denumerably) many zeros is zero; hence the net advantage of a *set* of bets at fair odds is zero; and, finally, (iii) if someone can, on the basis of an examination of the odds alone, be *assured* of a positive net gain or loss from simultaneous bets at those odds, then the net advantage in betting at those odds cannot be zero. From (i), (ii), and (iii) we conclude that assurance of a net gain or loss from finitely many simultaneous bets, from a knowledge of the odds alone, implies that the odds in each bet cannot all have been fair.

Let us now prove our result. To make the argument go as smoothly as possible we shall assume that bets with odds $\frac{p}{(1 - p)}$ on h, where p is the betting quotient, have the following canonical form. There will be two bettors, a bettor on h and a bettor against h, and a *stake S*, which we can take to be so many units of currency, where those units are assumed to be indefinitely divisible. The truth-value, true or false, of h is assumed to be capable of being settled to the agreement of both bettors, and when so settled a proportion of S is to be exchanged between the bettors thus: if h is determined to be true, then the bettor on receives $S(1 - p)$ from the bettor against. If h is decided to be false, then the bettor against receives pS from the bettor on. The payoff conditions for the bettor on look like this, therefore, where T stands for 'true' and F for 'false'

h	Payoff
T	$S(1 - p)$
F	$-pS$

This can be brought into what for most readers is the more familiar form of a bet at odds $\dfrac{p}{(1 - p)}$ by writing $Q = pS$. Then we obtain the conditions

h	Payoff
T	$Q\dfrac{(1 - p)}{p}$
F	$-Q$

where Q is the amount forfeited if h is agreed to be false, and Q multiplied by the reciprocal of the odds is the amount won.

The proof, which owes much to Skyrms (1977, Ch. VI), now proceeds as follows. Consider, first of all, the set consisting just of a single betting quotient $p = P(h)$ on a hypothesis h, and suppose that you violate axiom 1 of the probability calculus by setting $p < 0$. We have simply assumed so far that the minimum value of odds is 0: it is obvious, however, from the payoff diagrams above, that the negative odds obtained from making p negative would ensure that were anyone to be a bettor against, then they would make a sure loss, since then both that individual's entries, for T and F, would be negative. In other words, odds based on p, where $p < 0$, are demonstrably unfair, in that a positive advantage is possessed by one side of a bet on h at those odds.

Next, consider the set consisting just of a betting quotient $p = P(t)$ on a tautology t, and suppose that you violate the axiom which asserts that $p = 1$. There are two possibilities (a) $p < 1$, and (b) $p > 1$. In case (a) a bettor on h would necessarily gain $S(1 - p)$ from the bettor against, and in case (b) would necessarily lose $S(p - 1)$ to the bettor against. Neither, consequently, of these betting quotients can consistently be regarded as determining fair odds: only $p = 1$ does.

Thus axioms 1 and 2 of the probability calculus are necessary conditions for fairness: were these axioms not satisfied, then as the reasoning above shows, anybody with the equivalent degrees of belief, measured in the way we have measured them, would be classifying certain types of bet as fair which are demonstrably not.

Now consider the set $\{p_1,\ p_2,\ p_3\}$ where $p_1 = P(h_1)$, $p_2 = P(h_2)$, and $p_3 = P(h_1 \ \mathbf{v} \ h_2)$, and where h_1 and h_2 are mutually exclusive. Were someone to bet simultaneously on h_1 and h_2 at the corresponding odds $\dfrac{p_1}{(1 - p_1)}$ and $\dfrac{p_2}{(1 - p_2)}$, equal stakes S being placed on the bets, then their net payoffs would look like this:

h_1	h_2	Net Payoff
T	F	$S(1 - p_1) - p_2 S$
F	T	$-p_1 S + S(1 - p_2)$
F	F	$-p_1 S - p_2 S$

where, as before, T stands for 'true' and F for 'false', and the right-hand sums are your net gains corresponding to each of the three possible states of affairs. But the first two of these sums are identical; in other words, your gain if h_1 is true or h_2 is true is $S[1 - (p_1 + p_2)]$, and your loss if both hypotheses are false is $-S(p_1 + p_2)$. These are clearly the payoffs of a bet on $h_1 \ \mathbf{v} \ h_2$, with the same stake S, but at the odds $\dfrac{q}{(1 - q)}$, where $q = p_1 + p_2$. Thus the betting quotients p_1 and p_2 uniquely determine, in the context of simultaneous bets on h_1 and h_2, a betting quotient, $p_1 + p_2$, on the disjunction $h_1 \ \mathbf{v} \ h_2$. Were p_3 not to equal $p_1 + p_2$, then it follows that simultaneous bets on h_1, h_2, and $h_1 \ \mathbf{v} \ h_2$, at the odds determined respectively by p_1, p_2, and p_3, would entail two distinct odds being given on $h_1 \ \mathbf{v} \ h_2$.

The situation is thus analogous to an inconsistent set of sentences: the attempt to assign them all the truth-value 'true' results in one of the sentences being assigned different truth-values. In the present case the assignment is not of truth-values, but betting quotients, but the inconsistency is no less. Moreover, for anyone betting indifferently on or against a hypothesis at two distinct odds could be made to lose (or gain) a positive sum. This is easy to see: if p and q are distinct betting quotients on a hypothesis a, then betting on a at odds $\frac{p}{(1-p)}$ and against a at odds $\frac{q}{(1-q)}$ with the same stake S yields a positive net gain or loss as $p < q$ or $p > q$ respectively (for if a is true, the net gain is $S(1-p) - S(1-q)$; while if a is false, the net gain is $Sq - Sp$; and both these quantities are equal to $S(q-p)$). Thus were anyone to bet indifferently on or against h_1, h_2, and h_3 at odds p_1, p_2, and p_3, where $p_3 \neq p_1 + p_2$, then they could be made to suffer a net gain or loss, and this could be known in advance of any actual betting. Hence the betting quotients cannot all be fair, that is, give zero advantage to each side of each of the bets. For all of p_1, p_2, and p_3 to be fair, therefore, requires that $p_3 = p_1 + p_2$.

Before we turn to the remaining axiom, that of conditional probability, we shall show that the same argument, mutatis mutandis, requires not merely finite but countable additivity. Consider a lottery in which a natural number is to be named by some 'random' procedure. h_i is the hypothesis that the number i is selected. Let h be the hypothesis that an even number is selected, and h_{2j} the hypothesis that the particular even number $2j$ is selected. Then h is true just in case h_{2i} is true for some i. We should note at the outset that it follows almost immediately from the ordinary finite additivity property of the p_i that your associated degrees of belief p_i sum to a limiting value less than or equal to your degree of belief in h.

Suppose that the stake S is placed on each of the h_{2i}, on which you were to bet at the odds $\frac{p_{2i}}{(1-p_{2i})}$ respectively. If h_{2j} is true, then your net gain would be $S(-p_2 - p_4 - \ldots + (1 - p_{2j}) - \ldots) = S[1 - (p_2 + \ldots + p_{2n} + \ldots)]$, which is well defined and positive since, as we observed, the sum of the p_i converges to a value of at most 1. This gain is also independent of j. If h is false, then you would lose the quantity $S(p_2 + \ldots + p_{2n} + \ldots)$. So you would in effect be betting on h at the odds $\frac{q}{(1-q)}$ where $q = (p_2 + \ldots + p_{2n} + \ldots)$. To be consistent, then, your degree of belief in h must equal q.

There are, however, vigorous critics of the thesis that subjective probabilities are countably additive. De Finetti, for example, has produced many alleged counter-arguments. To reassure the reader, however, that we are not dismissing out of hand these objections from someone whose authority is certainly not to be considered lightly, let us consider briefly one of the most seductive of these counter-arguments. This considers the example of an integer being chosen 'at random', in which case it might seem natural to require a uniform, zero, degree of belief in each integer being selected. This is quite consistent with finite additivity, but not countable additivity (because then the sum of these degrees of belief would have to be 1). But as Spielman (1977) points out, it is not at all clear what selecting an integer at random could possibly amount to: any actual process would inevitably be biased toward the 'front end' of the sequence of positive integers, and so there is in reality little force in de Finetti's counter-example. Let us now move on to consider the remaining probability axiom, that of conditional probabilities.

d.3 Conditional Betting Quotients

It only remains now to consider how the axiom 4, of conditional probabilities, is satisfied. First we have to fix an interpretation of a conditional probability within this theory of one's assessment of fair odds. This is quite easy: if $P(a)$ is interpreted as what you think is a fair betting quotient on a, then the conditional probability $P(a \mid b)$ is naturally interpreted as the betting quotient on a which you think would be fair were you to come to know the truth of b and nothing stronger, where b is assumed to be logically independent of your background information. $P(a \mid b)$ signifies, in other words, that you would deem fair a *conditional bet* on a at odds $\dfrac{p}{(1-p)}$, where $p = P(a \mid b)$.

A bet is a conditional bet on a given b at odds $\dfrac{p}{(1-p)}$ if it goes ahead at those odds on receipt of an acceptable affidavit of b's truth and is called off on receipt of an acceptable affidavit of b's falsity. $P(a \mid b)$ is called your conditional degree of belief in a, given b. What we shall now show is that by setting appropriate stakes on a and $a \, \& \, b$, simultaneous bets on those two statements are equivalent to a bet on a, conditional on the truth of b, and that any odds placed on a and $a \, \& \, b$ can therefore be made to determine the odds for a conditional bet on a given b.

A bet on a conditional on the truth of b, at odds $\dfrac{p}{(1-p)}$ and with stake S, has the payoff conditions

a	b	Payoff
T	T	$S(1 - p)$
F	T	$-pS$
	F	0

Now suppose that you were to bet on a & b at odds $\dfrac{q}{(1 - q)}$ and against b at odds $\dfrac{r}{(1 - r)}$, and that the amounts staked were V and S respectively. Your net gains, depending on what happens are therefore as follows

a & b	b	Net Payoff
T	T	$V(1 - q) - S(1 - r)$
F	T	$Vq - S(1 - r)$
F	F	$-Vq + Sr$

But now let $V = r$ and $S = q$. Then your net payoffs are clearly

a & b	b	Net Payoff
T	T	$r - q$
F	T	$-q$
F	F	0

which is obvoiusly equivalent to

a	b	Net Payoff
T	T	$r\left(\dfrac{1 - q}{r}\right)$
F	T	$-r\left(\dfrac{q}{r}\right)$
	F	0

This is clearly the payoff matrix of a conditional bet on a given b with conditional betting quotient $\frac{q}{r}$, *i.e.* the ratios of the betting quotients q on a & b and r on b. As with two mutually exclusive hypotheses, therefore, simultaneous bets with appropriate stakes will also determine a further bet—in this case, a conditional one. Hence if the conditional betting quotient $P(a|b)$ you actually state were to differ from $\frac{q}{r}$, you would implicitly be assigning different odds to the same (conditional) hypothesis; and, as we saw earlier, anybody betting indifferently on or against a hypothesis at different odds can be made to suffer a net gain or loss independently of the truth-values of the hypothesis. In order that the three betting quotients can consistently be regarded as being fair, therefore, the betting quotient on the conditional prediction must be related to the two unconditional betting quotients according to axiom 4 of the probability calculus.

We have now completed the proof that if a set of betting quotients do not satisfy the probability calculus, then they certainly cannot all be fair. As we pointed out earlier, this result is independent of any formal characterisation of fairness of odds beyond the stipulation that they confer no advantage to either side of a bet at those odds. The result follows because the sum of a finite number of zeros is zero, so that an assured positive net gain or loss is not consistent with a zero advantage to each bet. We shall now, to round off the discussion, consider a particular method, used since the eighteenth century, of computing the value of the advantage to taking a particular side in a bet.

d.4 Fair Odds and Zero Expectations

Laplace (1820, p. 20) defined the *advantage* to taking a given side in a wager to be the expected value, relative to a probability distribution p, $1 - p$ which allegedly defines the rational degree of belief to be entertained in h and $\sim h$. Thus advantage is calculated in the same units as the stake S, and so can be subjected to straightforward arithmetical operations, like taking sums of separate advantages. Carnap, his twentieth-century successor, calls the same expected value the "estimated gain" (1950, p. 170) and a bet fair just when the "estimated gain" is zero, where the expectation is computed relative to an appropriate Carnapian c-function. It is an easy task to show that this expectation is zero when and only when the odds are equal to $\dfrac{p}{(1-p)}$, or $\dfrac{c}{(1-c)}$ in the Carnapian system. For suppose that you will win V if h is true, and lose R if not. These are the payoff conditions of betting on h at odds $\dfrac{R}{V}$, and the bet itself can therefore be represented by a two-valued random variable X taking the value V with probability p, and $-R$ with probability $1 - p$. Its expected value is therefore $Vp - R(1 - p)$ relative to that probability distribution, and this quantity is obviously zero just when the odds $\dfrac{R}{V}$ are equal to $\dfrac{p}{(1-p)}$. But as we saw earlier in this chapter, the sort of probability distribution envisaged by Laplace and his twentieth-century successors presupposes the sort of a priori probability measure whose existence there is every reason to doubt. However, if we use subjective probabilities instead and explicitly define the advantage to be the expected value, relative to these probabilities, of betting at those odds, then we deduce as a theorem that the subjective advantage to betting at odds $\dfrac{R}{V}$ is zero if and only if those odds are equal to the ratio, $\dfrac{p}{(1-p)}$, of your degrees of belief in h and $\sim h$ respectively.

We have, in other words, found a mathematical representation of the informal notion of advantage which yields as a consequence the desired result that degrees of belief are subjectively fair betting quotients: the informal criterion, that a bet is subjectively fair when you feel that no advantage accrues to either side of the bet at the associated odds, becomes translated into the formal result that the expected value, relative to your subjective probability distribution p, $1 - p$, of taking either side of a bet at odds $\dfrac{p}{(1-p)}$, is zero. This result does not of course prove anything substantially new; it merely shows that the informal notion of subjective fairness can be, to use

Carnapian terminology, given a formal explication which preserves all the desired consequences.

d.5 Fairness and Consistency

We have laid a foundation for a theory of degrees of belief, whose methodological consequences we shall explore in the subsequent chapters, in the notion of subjectively fair odds. But is there any sense in which odds other than those on tautologies and contradictions may be objectively fair? One candidate for a criterion of objective fairness was, as we have seen, having zero expectation relative to a 'logical' probability distribution of the type Laplace, Keynes, and Carnap tried to define. We have seen that their attempts foundered on the rock of pure arbitrariness. However, there is famously an alternative criterion: odds are fair when they are determined by the real *physical* probabilities of the events concerned, where those probabilities exist. We believe that, with certain qualifications, this claim is true, and indeed we shall base our theory of statistical inference on it. But any argument for that thesis must await a discussion of the notion of physical probability itself, a notion which, as we shall see, is fraught with difficulties. We shall take up that discussion again in Chapter 9.

We conclude this section by introducing a useful piece of terminology. Ramsey (1931) used the term "consistent" to characterise degrees of belief having the formal structure of probabilities. De Finetti used the term "coherent", and it is de Finetti's terminology that is usually employed today. We prefer Ramsey's, since, as we have shown in the previous section, there is an implicit contradiction in having degrees of belief—in maintaining that qua betting quotients those numbers determine fair odds—such that bets at the associated odds are demonstrably unfair. We shall now show that consistency considerations give us a further important constraint on degrees of belief, and one which will be the foundation of the theory of Bayesian inference: the Principle of Conditionalisation.

d.6 Conditional Probabilities and Changing Beliefs

Suppose that h is some hypothesis and e some evidence bearing on the truth of h; for the sake of illustration suppose that e is the outcome of a test of h. We have not yet explained why we are supposing that the difference between the conditional probability $P(h \mid e)$ and the absolute probability $P(h)$ should represent the extent to which your degree of belief in h changes on receipt of e. This is not obvious, particularly in view of the interpretation of the conditional probability $P(h \mid e)$ as a conditional degree of belief, and therefore an opinion given apparently *before* the receipt of the information e. This fact has led some people (the first was probably Hacking, 1967) to claim that the Bayesian theory offers no justification for supposing that Bayes's Theorem establishes a relation between pre- and post-test probabilities—for *all* the probabilities appearing there are pre-test probabilities. What it is necessary to show to exploit Bayes's Theorem in this way is that $P'(h) = P(h \mid e)$, where $P'(h)$ is your degree of belief in h after receipt of e, and this is an independent principle, which has come to be called the Principle of Conditionalisation. It is a principle adopted by Bayesians, but one which, according to critics of the Bayesian theory, lacks any real justification. Thus Kyburg, following Hacking, asserts that "there is nothing in the [Bayesian] theory that says that a person should *change* his beliefs in response to evidence in accordance with Bayes's theorem" (1983, p. 95).

To see that this conclusion is incorrect, however, is very straightforward. If, as will be assumed, the background information relative to which $P'(h)$ is defined differs from that to which $P(h \mid e)$, $P(h)$, $P(e)$, etc., are relativised only by the addition of e, the principle follows immediately; for $P(h \mid e)$ is, as far as you are concerned, just what the fair betting-quotient would be on h were e to be accepted as true. Hence from the knowledge that e is true you should infer (and it is an inference endorsed by the standard analyses of subjunctive conditionals) that the fair betting quotient on h is equal to $P(h \mid e)$. But the fair betting quotient on h after e is known is by definition $P'(h)$.

■ e INTERVAL-VALUED PROBABILITIES

We promised that we would return to discuss hypotheses and data sets where it is unrealistic to suppose that one would have point-valued degrees of belief. To borrow an example from Suppes (1981, p. 41): if we consider the question of whether it will rain at some specified time in Fiji, we can certainly suggest a value $\mathbb{O}_1$ such that odds less than $\mathbb{O}_1$ on that hypothesis are in our opinion unrealistically low, and we can also suggest odds $\mathbb{O}_2$ such that odds greater than $\mathbb{O}_2$ against rain are in our opinion unrealistically high. But we should have to concede also there will be an intermediate interval of odds between which we are unable to discriminate and whose precise extent, if indeed it has a determinate length at all, reflects the extent of our ignorance. If we are totally ignorant, the interval would presumably be the entire unit interval. Thus the typical indefiniteness of one's knowledge would, it seems, be more faithfully reflected by an interval-valued function which only in certain cases takes degenerate intervals, or points, as values.

Interval-valued probability functions have been fairly thoroughly investigated, and they form the subject of the theory of what are called upper and lower probabilities. The terminology was introduced by Koopman (1940) and the theory has since been developed from a variety of starting points by Good (1962), Smith (1961), Dempster (1968), Williams (1976), and others. Upper and lower probabilities are usually symbolised by P^* and P_* respectively, and they have some simple formal properties: for example, P^* is never less than P_*; and P_* is super-additive and P^* sub-additive, meaning that if $h_1 \vdash \sim h_2$, then $P_*(h_1 \vee h_2) \geq P_*(h_1) + P_*(h_2)$, the inequality going the other way for P^*. Where $P^*(h) = P_*(h)$, then that quantity is simply called the probability of h; probabilities so defined obey the usual rules of the probability calculus.

The theory of upper and lower probabilities is considerably weaker than that of point-probabilities: it replaces, as we can see from the relations above, equalities by inequalities in general (though not invariably). This means in particular that many of the striking relationships between prior and posterior belief relative to appropriate types of evidence become difficult if not impossible to derive. In the point-probability model, on the other hand, they are obtained very simply, and due allowance can then be made, if necessary, for any imprecision in the prior distributions. Moreover, the cost incurred by departing from the point-probability model does not really, despite the appearance of greater fidelity to the phenomenon of inexact belief states, bring commensurate rewards of greater realism. Just as there are probably no exact point-valued degrees of belief, so there are probably also no exact point-valued bounds to the range of imprecision of the belief. In which case there is nothing to be lost and much to be gained in going for the stronger theory, which we can regard as describing an ideal for whose realisation we strive, and which is satisfied more or less approximately in many cases. One obvious example is that of drawing a card from a deck of cards after it has been thoroughly shuffled: most people would, in these circumstances, nominate $\frac{1}{52}$ as the fair betting quotient on drawing any specified card.

We should also stress two facts that seem frequently to get overlooked in the discussion of the value of a point-probability model. One is that there is nothing in the theory we have put forward which asserts that people actually do have point-valued degrees of belief. That theory is quite compatible with people's beliefs being as vague as you like: it merely states that if they were to be point-valued, and consistent, then they would formally be probabilities. The second fact is that the same divergence between mathematical theory and reality is equally characteristic of the measurement of all those *physical* magnitudes, where people are quite happy to invoke real number values. We suffer no attack of conscience in applying the theory of real-number arithmetic to compute areas of rooms, volumes of solids, or what have you, even though, according to current physical theory, it is quite false to say that any of these magnitudes is measured by a real number, or even that they lie in a precisely specified non-degenerate interval of real numbers. Real number theory, applied to the measurement of physical quantities, is so widely used because, as an idealisation which often gives sufficiently accurate results within the ranges of imprecision in which we work, it is indispensable. Yet for some reason people feel that because our degrees of belief are accessible only within greater or smaller intervals, that is, by itself, a sufficiently serious objection to the use of a model invoking real-number probabilities that that model ought to be abandoned. This is not the case.

But we have already started to stray too much into the discussion, which we have reserved for the last chapter of this book, of objections to our theory and the extent to which they are valid. Let us regard the issue of point- versus interval-valued probabilities as now sufficiently discussed and take the proof of the pudding to be in the eating: we feel confident that by the end of the book the reader will have become convinced of the utility of the point-probability model. We shall now close this chapter with a brief conspectus of other arguments for supposing that subjective uncertainty, *where* it is considered to be measurable at all by means of real numbers, is measurable by functions which obey the probability calculus.

■ f OTHER ARGUMENTS FOR THE PROBABILITY CALCULUS

f.1 The Standard Dutch Book Argument

Any reader of books or articles on philosophical probability will be struck by the variety of arguments produced for the thesis that strength of belief can be measured numerically and that one such measure satisfies the axioms of the probability calculus. We have ourselves produced an argument for this conclusion, where we identified degrees of belief with subjectively fair betting-quotients. This argument exploits a purely arithmetical result proved independently by Ramsey (1931) and de Finetti (1937): that if people were to bet at odds derived from betting quotients which do not satisfy the probability calculus, and the side of the bet they took and the size of the stake could be dictated by the other bettor, the latter could in principle win a positive sum come what may. In gambling parlance the victims would have left themselves open to a Dutch Book. Let us from now on call this purely arithmetical result the Dutch Book Argument.

We used the Dutch Book Argument to show that a set of betting quotients which failed to satisfy the probability calculus could not all be fair. We then justified invoking the constraints imposed by the probability calculus on degrees of belief by identifying the latter as subjectively fair betting-quotients. Another way in which the Dutch Book Argument has been used to justify obedience to the probability calculus is altogether more dubious, however. This proceeds from the postulate that to possess a degree of belief p in h is actually to be prepared to bet indifferently on or against h at odds $\dfrac{p}{(1 - p)}$, so long as the stakes are kept small. The Dutch Book Argument then, clearly, entails that if your degrees of belief, so characterised, do not satisfy the probability calculus, then there are positive and negative stakes (positive stakes mean you bet on, negative stakes mean you bet against) which you would accept in bets at the odds determined by your degrees of belief and which, once accepted, would cause you to lose money come what may. It is (plausibly) assumed that you would not voluntarily retain such a system of degrees of belief once their vulnerability to a Dutch Book had been brought to your attention, and it is then concluded that obedience to the probability calculus is a principle of *economic* rationality.

The trouble with this use of the Dutch Book Argument, which critics have not been slow to point out, is that the postulate that degrees of belief entail willingness to bet at the odds based on them is vulnerable to some rather obvious objections. One is that there are hypotheses for which the wise choice of odds bears no relation to your *real* degree of belief: thus if h is an unrestricted universal hypothesis over an infinite domain, then it is clear that while it may in certain circumstances be possible to falsify h, it is not possible to verify it. Thus the only sensible *practical* betting quotient to nominate on h is 0; for you could never gain anything if your betting quotient was positive and h was true, while you would lose if h turned out to be false. Yet you might well believe that h stands a nonzero chance of being true.

Indeed, as we observed earlier, there may be all sorts of reasons, and even possibly none at all, why one might be unwilling to bet at odds commensurate with one's actual degree of belief. Most people, for example, are induced to bet only at odds which they think are advantageous to them. Even if the stakes are kept very small, then as Ramsey observed (1931, p. 176), it is difficult to see why you should bother to be too realistic when announcing odds at all. It is the problematic nature of the postulate which prompted C. A. B. Smith (1961) to abandon it and assume merely that there is a minimum of the odds you would accept on a given hypothesis, and a maximum at which you would take the other side of the bet, and that these usually will not coincide (he then used Dutch Book considerations to show that these maxima and minima determine upper and lower probabilities).

De Finetti (1937, p. 102) and later Mellor (1971, p. 37) attempt to evade the difficulty by introducing an element of compulsion: you are now compelled to name betting quotients, with your coercer free to appoint (i) the hypotheses to be bet on, (ii) which side of each of the bets you will take, and (iii) the stakes. The quotients you pick are then identified with your degrees of belief, and the Dutch Book Argument is invoked to show that they should satisfy the probability calculus. But the introduction of compulsion still leaves the identification of genuine degrees of belief with the betting quotients thus elicited open to serious, and in our opinion insuperable, objections. For example, there still remains the problem of hypotheses which are only one-way decidable: there is still no justification for naming other than zero odds on these. In addition, there is now no reason why the presumptive degrees of belief should obey the probability calculus. An example will make this clear. Suppose that somebody were to inflict on you the most hideous torture were you to deny that $2 + 2 = 5$. It would be perfectly rational of you *in those circumstances* to agree that $2 + 2 = 5$, but it certainly would not necessarily be rational to extend your agreement outside them; indeed, there is no reason at all why you should. The same holds for the forced-betting situation we are being asked to contemplate; it is simply not a good reason to give that just because there are situations like the one in which penalties are attached to a set of inconsistent betting quotients, those betting quotients should *in general* obey probabilistic constraints.

f.2 Scoring Rules

Let I_h be a function which takes the value 1 for those 'possible worlds' in which h is true, and 0 for those in which it is false (if h describes a generic outcome of some stochastic experiment E, then the 'possible worlds' on which I_h is defined can be regarded simply as the distinct basic outcomes of E). I_h is called the indicator function of h. Let p be a real number between 0 and 1 inclusive. The payoffs of a bet on h at odds $\dfrac{p}{(1 - p)}$ with stake S can be represented by the values of $S(I_h - p)$, for these are $S(1 - p)$ when h is true, and $-Sp$ when h is false. By allowing S to take negative values, the payoff function has the same form whichever side of the bet you take: if you are betting on h, S is positive; if against, negative. Where p represents your degree of belief in h, this function is a type of what has come to be called scoring rule, or function $f(I_h, p)$ which for each h assigns a payoff depending on the values of I_h and p.

Usually scoring rules are defined not in terms of gains but of penalties $f(I_h, p)$ which you incur when you announce p, and h is evaluated. De Finetti (1974, p. 87) introduces the quadratic scoring rule (again, in a rather more general form) $L(I_h, p)$, which exacts a penalty proportional to $(I_h - p)^2$. Lindley (1982) contains a very striking result about (penalising) scoring rules. He defines a set of degrees of belief ('uncertainties') $p_1, \ldots, p_n$ to be admissible if for every other set $p'_1, \ldots, p'_n$ there are values of the indicator variables I_{h_i} such that the sum of the scores $f(I_{h_i}, p_i)$ is less than the sum of the scores $f(I_{h_i}, p'_i)$, and for all values of the I_{h_i} the sum of the $f(I_{h_i}, p_i)$ never exceeds that of the $f(I_{h_i}, p'_i)$. Lindley shows that if $p_1, \ldots, p_n$ are admissible, then there is a continuous mapping of $p_1, \ldots, p_n$ into the closed unit interval which satisfies axioms 1–4 of the probability calculus. In other words, if your degrees of belief are such that they could not be replaced by a different set which would give a lower net loss for some events, and is never higher on any, then they are either directly probabilities or can be rescaled to become probabilities.

C. A. B. Smith, in the discussion to Lindley's paper, shows that by replacing the uniform penalties in Lindley's approach by positive and negative rewards, the scoring rule in that theory can be transformed into the standard form of bets in which the agent indifferently takes either side of the bets at the given odds determined by his degree of belief. But as Smith points out, this is just the unrealistically strong assumption which we pointed out earlier vitiates the force of the classic use of the Dutch Book Argument.

f.3 The Cox-Good-Lucas Argument

Cox (1961), Good (1950, Appendix III), and Lucas (1970) propose variants of a quite different type of argument for supposing that subjective probabilities are technically probabilities in the sense of the probability calculus. They each show that if a real-valued function f satisfies a few very general and plausible desiderata for a measure of reasonable degree of belief, then f can be transformed ('regraduated') by a suitably increasing function $Q(f)$ into a conditional probability function. For example, all that Cox requires is that

$$f(c \,\&\, b \mid a) = G[f(c \mid b \,\&\, a), f(b \mid a)]$$

and that

$$f(\sim b \mid a) = H(f(b \mid a))$$

for some functions G, H satisfying certain differentiability conditions. He is then able to show that

$$0 \le Q(f) \le 1$$

and that

$$Q[f(c \& b \mid a)] = Q(f(c \mid b \& a)) \cdot Q(f(b \mid a))$$

and

$$Q[f(\sim b \mid a)] = 1 - Qf(b \mid a).$$

From these properties of the transform Q it is not difficult to show that it satisfies our axioms 1–4 of the probability calculus.

f.4 Savage's Argument

Another argument for supposing that subjective probabilities should obey the standard (finitely additive) probability axioms is due to Savage (1954), who follows Ramsey's lead in developing simultaneously a theory of subjective probability and a theory of utility in the context of a general theory of preferences between acts. If one considers specifically those acts (like simple bets) which involve receipt of a prize if a specified event occurs, then Savage shows how from a linear (reflexive) preference ordering between these acts we may derive an ordering $\leq$ of those events themselves, this latter ordering called by him a qualitative personal probability. He then shows that subject to certain additional conditions being satisfied, there is a unique probability function P such that $a \leq b$ if and only if $P(a) \leq P(b)$, where the latter $\leq$ is the usual ordering of the real numbers.

■ g CONCLUSION

These arguments vary in their persuasiveness; ingenious though they are, none seems to provide any conclusive reason for asserting that degrees of belief ought to obey the probability calculus. They all depend on conditions which it may seem counterintuitive to deny, but which can nevertheless certainly be denied without inconsistency. However, they are far from valueless. It is a striking fact that, starting from often apparently very different assumptions, all plausible in their own way, so many arguments lead directly to the probability calculus. The latter seems, in other words, to be a sort of invariant of different ways of defining uncertainty, or as Lindley puts it, "inevitable", meaning that the choice of any plausible way of mathematically measuring uncertainty will lead to it. This convergence of arguments has a powerful cumulative effect and increases our conviction that the probability calculus corresponds to some quite objective feature of subjective uncertainty. The situation is rather analogous to that in the mathematical theory of effective computability, where a number of apparently distinct ways of characterising that notion turn out to define exactly the same class of functions, the so-called partial recursive functions. But enough, we feel, has been said about why degrees of belief should be formally probabilities; let us now see what methodological consequences flow from assuming that they are.

PART II

Bayesian Induction: Deterministic Theories

Philosophers of science have traditionally concentrated attention primarily on deterministic hypotheses, leaving statisticians to discuss the methods by which statistical or non-deterministic theories should be assessed. Accordingly, a large part of what would more naturally be regarded as philosophy of science is normally treated as a branch of statistics, going under the heading 'statistical inference'. So it is not surprising that philosophers and statisticians have developed distinct methods for their different purposes. We shall, in parts II and III of this book, follow the tradition of treating deterministic and statistical theories separately. However, as will become apparent, we regard this separation as artificial and shall, in the course of the book, expound the unified treatment of scientific method afforded by Bayesian principles.

■ CHAPTER 4
Bayesian Versus Non-Bayesian Approaches

In this chapter we shall consider how, by attributing positive probabilities to hypotheses in the manner described in Chapter 2, one can account for many of the characteristic features of scientific practice, particularly as they relate to deterministic theories.

■ a THE BAYESIAN NOTION OF CONFIRMATION

Information gathered in the course of observation is often considered to have a bearing on the acceptability of a theory or hypothesis (we use the terms interchangeably), either by confirming it or by disconfirming it. Such information may either derive from casual observation or, more commonly, from experiments deliberately contrived in the hope of obtaining relevant evidence. The idea that evidence may count for or against a theory, or be neutral towards it, is a central feature of scientific inference, and the Bayesian account will clearly need to start with a suitable interpretation of these concepts.

Fortunately, there is a suitable and very natural interpretation, for if $P(h)$ measures your belief in a hypothesis when you do not know the evidence e, and $P(h \mid e)$ is the corresponding measure when you do, e surely confirms h when the latter exceeds the former. So we shall take the following as our definitions:

e **confirms or supports** h when $P(h \mid e) > P(h)$

e **disconfirms or undermines** h when $P(h \mid e) < P(h)$

e **is neutral with respect to** h when $P(h \mid e) = P(h)$

One might reasonably take $P(h \mid e) - P(h)$ as measuring the degree of e's support for h, though other measures have been suggested (e.g., Good, 1950). Disagreements on this score will not be controversial in this book. We shall refer, in the usual way, to $P(h)$ as 'the prior probability of h' and to $P(h \mid e)$ as h's 'posterior probability' relative to, or in the light of, e. The reasons for this terminology are obvious, but it ought to be noted that the terms have a meaning only in relation to evidence: as Lindley (1970, p. 38) put it, "Today's posterior distribution is tomorrow's prior". It should be remembered too that all the probabilities are evaluated in relation to accepted background knowledge.

■ b THE APPLICATION OF BAYES'S THEOREM

Bayes's Theorem relates the posterior probability of a hypothesis, $P(h \mid e)$, to the terms $P(h)$, $P(e \mid h)$, and $P(e)$. Hence, knowing the values of these last three terms, it is possible to determine whether e confirms h, and, more importantly, to calculate $P(h \mid e)$. In practice, of course, the various probabilities may only be known rather imprecisely; we shall have more to say about this practical aspect of the question later.

The dependence of the posterior probability on the three terms referred to above is reflected in three striking phenomena of scientific inference. First, other things being equal, the extent to which evidence e confirms a hypothesis h increases with the likelihood of h on e, that is to say, with $P(e \mid h)$. At one extreme, where e refutes h, $P(e \mid h) = 0$; hence, disconfirmation is at a maximum. The greatest confirmation is produced, for a given $P(e)$, when $P(e \mid h) = 1$, which will be met in practice when h logically entails e. Statistical hypotheses, which will be dealt with in parts III and IV of this book, are more substantially confirmed the higher the value of $P(e \mid h)$.

Secondly, the posterior probability of a hypothesis depends on its prior probability, a dependence sometimes discernible in scientific attitudes to ad hoc hypotheses and in frequently expressed preferences for the simpler of two hypotheses. As we shall see, scientists always discriminate, in advance of any experimentation, between theories they regard as more or less credible and, so, worthy of attention and others.

Thirdly, the power of e to confirm h depends on $P(e)$, that is to say, on the probability of e when it is not assumed that h is true (which, of course, is not the same as assuming h to be false). This dependence is reflected in the scientific intuition that the more surprising the evidence, the greater its confirming power. However, $P(e) = P(e \mid h)P(h) + P(e \mid \sim h)P(\sim h)$, as we showed in Chapter 2, section **e,** so that really, the posterior probability of h depends on the three basic quantities $P(h)$, $P(e \mid h)$, and $P(e \mid \sim h)$.

We shall deal in greater detail with each of these facets of inductive reasoning in the course of this chapter.

■ c FALSIFYING HYPOTHESES

A characteristic pattern of scientific inference is the refutation of a theory, when one of the theory's empirical consequences has been shown to be false in an experiment. As we saw, this kind of reasoning, with its straightforward and unimpeachable logical structure, exercised such an influence on Popper that he made it into the centrepiece of his scientific philosophy.

Although the Bayesian approach was not conceived specifically with this aspect of scientific reasoning in view, it has a ready explanation for it. The explanation relies on the fact that if, relative to background knowledge, a hypothesis h entails a consequence e, then (relative to the same background knowledge) $P(h \mid \sim e) = 0$. Interpreted in the Bayesian fashion, this means that h is maximally disconfirmed when it is refuted. Moreover, it can be shown that, as we should expect, once a theory is refuted, no further evidence can confirm it, unless the evidence or some part of the background assumptions are revoked. (This is simply proved: if h entails e, then h & $\sim e$ is a contradiction, so $P(h$ & $\sim e) = 0$, whence $P(h \mid e) = 0$. And if f is some further datum, then since h & $\sim e$ & f is also a contradiction, the same argument shows that $P(h \mid \sim e$ & $f) = 0$.)

■ d CHECKING A CONSEQUENCE

A standard method of investigating a deterministic hypothesis is to draw out some of its logical consequences, relative to some stock of background theories, and check whether they are true or not. For instance, the General Theory of Relativity was confirmed by establishing that light is deflected when it passes near the sun, as the theory predicts. It is easy to show, by means of Bayes's Theorem, why and under what circumstances a theory is confirmed by its consequences.

If h entails e, then, as may be simply shown, $P(e \mid h) = 1$. Hence, from Bayes's Theorem: $P(h \mid e) = \dfrac{P(h)}{P(e)}$. Thus, if $0 < P(e) < 1$, and if $P(h) > 0$, then $P(h \mid e) > P(h)$. It follows that any evidence whose probability is neither of the extreme values must confirm every hypothesis with a non-zero probability of which it is a logical consequence.

Succeeding confirmations must eventually diminish in force, for the theory has an upper limit of probability, beyond which no amount of evidence can push it. And as the theory becomes more probable with the accumulation of evidence, further consequences of the theory acquire a greater likelihood of being true, and thus a smaller power to confirm. All this follows from Bayes's Theorem. Suppose $e_1, e_2, \ldots e_n \ldots$ are consequences of h, which are found to be true. Then Bayes's Theorem asserts that

$$P(h \mid e_1 \,\&\, e_2 \ldots \&\, e_n) = \frac{P(h)}{P(e_1 \,\&\, e_2 \ldots \&\, e_n)}$$

Now

$$P(e_1 \,\&\, e_2 \ldots \&\, e_n) = P(e_1)P(e_2 \,\&\, \ldots \&\, e_n \mid e_1)$$

and

$$P(e_2 \,\&\, \ldots \&\, e_n \mid e_1) = P(e_2 \mid e_1)P(e_3 \,\&\, \ldots \&\, e_n \mid e_1 \,\&\, e_2)$$

Thus, in general,

$$P(e_1 \,\&\, e_2 \,\&\, \ldots \&\, e_n) =$$
$$P(e_1)P(e_2 \mid e_1) \ldots P(e_n \mid e_1 \,\&\, \ldots \&\, e_{n-1})$$

Hence,

$$P(h \mid e_1 \,\&\, e_2 \,\&\, \ldots \&\, e_n)$$
$$= \frac{P(h)}{P(e_1)P(e_2 \mid e_1) \ldots P(e_n \mid e_1 \,\&\, \ldots \&\, e_{n-1})}.$$

Provided $P(h) > 0$, the term $P(e_n \mid e_1 \,\&\, \ldots \&\, e_{n-1})$ must tend to 1. If it did not, the posterior probability of h would at some point exceed 1, which is impossible (Jeffreys, 1961, pp. 43–44). This explains why one would not continue to test a hypothesis indefinitely, though without more detailed information on the individual's belief-structure, in particular regarding the values of $P(e_n \mid e_1 \,\&\, \ldots \&\, e_{n-1})$, one could not predict the precise point beyond which further predictions of the hypothesis were sufficiently probable not to be worth examining.

Specific categories of a theory's consequences also have a restricted capacity to confirm (Urbach, 1981). Suppose h is the theory under discussion and that h_r is a substantial restriction of that theory. A substantial restriction of Newton's theory might, for example, express the idea that freely falling bodies near the earth descend with a constant acceleration or that the period and length of a pendulum are related by the familiar formula. Since h entails h_r, $P(h) \leq P(h_r)$ (*see* Chapter 2, section **e**), and if h_r is much less speculative than its progenitor, it will often be significantly more probable.

Now consider a series of predictions derived from h, but which also follow from h_r. These may then confirm both theories, their posterior probabilities being given by Bayes's Theorem, thus:

$$P(h \mid e_1 \, \& \, e_2 \ldots \& \, e_n) = \frac{P(h)}{P(e_1 \, \& \, e_2 \ldots \& \, e_n)}$$

and

$$P(h_r \mid e_1 \, \& \, e_2 \ldots \& \, e_n) = \frac{P(h_r)}{P(e_1 \, \& \, e_2 \ldots \& \, e_n)} \, .$$

Combining these two equations to eliminate the common denominator, one obtains

$$P(h \mid e_1 \, \& \, e_2 \ldots \& \, e_n) = \frac{P(h)}{P(h_r)} \times P(h_r \mid e_1 \, \& \, e_2 \ldots \& \, e_n).$$

Since the maximum value of the last probability term in this equation is 1, it follows that however many predictions of h_r are verified, the main theory, h, can never acquire a posterior probability in excess of $\frac{P(h)}{P(h_r)}$. Hence, the type of evidence characterised by entailment from h_r may well be limited in its capacity to confirm h. This explains the phenomenon that repetitions of an experiment often confirm a general theory only to a limited extent, for the predictions verified by means of a given kind of experiment (that is, an experiment designed to a specified pattern) do normally follow from and confirm a much restricted version of the predicting theory.

When an experiment's capacity to generate confirming evidence has been exhausted through repetition, further support would have to be sought from other experiments, moreover, experiments of *different kinds*. We have an intuitive grasp on the idea of diversity among experiments. For instance, measuring the melting point of oxygen on a Monday and on a Tuesday would be the same experiment, but would be different from determining the rate at which oxygen and hydrogen react to form water. Ascertaining this reaction rate under different temperature and pressure conditions would presumably also count as different experiments, though it seems natural to say in such cases that the differences are not so great.

Franklin and Howson (1984) characterised similarity amongst experiments in a way which does considerable justice to these intuitions. They considered two experiments, E and E', each capable, in principle, of being instantiated indefinitely and yielding, respectively, the outcomes e_1, e_2, ... and e'_1, e'_2, ... They suggested that E and E' are different just in case for all $m > m_o$, and for some m_o,

$$P(e_{m+1} \,|\, e_1 \,\&\, e_2 \ldots \,\&\, e_m) > P(e'_i \,|\, e_1 \,\&\, e_2 \ldots \,\&\, e_m)$$

and for all $n > n_o$, and for some n_o,

$$P(e'_{n+1} \,|\, e'_1 \,\&\, e'_2 \ldots \,\&\, e'_n) > P(e_j \,|\, e'_1 \,\&\, e'_2 \ldots \,\&\, e'_n).$$

What this condition states, in other words, is that beyond a certain number of repetitions of E (respectively E'), the probability of a further outcome of that experiment is greater than the probability of any outcome of E' (respectively E). When this is the case, Franklin and Howson say that E and E' are different. (There is clearly scope for extending this definition to cover the notion of degrees of difference between experiments.) The definition means that if all the experimental outcomes follow deductively from some hypothesis, and if one experiment had been performed sufficiently often, then that hypothesis would be more substantially confirmed by the outcome of a different experiment than by a further instance of the same one, which is what we set out to show. The closely related fact that a wide variety of data gives greater support to a hypothesis than an equally extensive collection of similar data will be discussed later in the chapter, under a different heading (section **j.5**).

The arguments and explanations in this section rely on the possibility that evidence already accumulated from an experiment may increase the probability of further performances of that experiment producing similar results. Such a possibility is contested by Popperians, who rule it out as an unacceptable deviation from a purely deductive pattern of reasoning. But in so doing, they appear to rule out any explanation for the fact, attested by every scientist, that by repeating some experiment, one eventually (usually quickly) exhausts its capacity to confirm a given hypothesis. Alan Musgrave (1975), however, thought the fact could be explained non-inductively, in a manner compatible with Popperian principles. He claimed that after a certain number of repetitions of an experiment, the scientist would form a generalisation to the effect that whenever the experiment is performed, it yields a similar result. Musgrave then suggested that the generalisation would be entered into 'background knowledge'. Relative to this newly augmented background knowledge, the experiment is certain to produce a similar result on its next performance. Musgrave then appealed to the principle that evidence confirms a hypothesis in proportion to the difference between its probability relative to the hypothesis together with background knowledge and its probability relative to background knowledge alone. (That is, in Popper's notation, confirmation is proportional to $P(e \mid h \& b) - P(e \mid b)$, where b is background knowledge.) Musgrave then inferred that even if the experiment did produce the expected result when next performed, the hypothesis would receive no new confirmation. Watkins (1984, p. 297) more recently concurred with this account.

A number of objections may be made against it, though. First, as we shall show in the next section, although it seems to be a fact and is an essential constituent of Bayesian reasoning, there is no basis in the Popperian methodology for confirmation to depend on the probability of the evidence; Popper simply invoked the principle ad hoc. Secondly, Musgrave's suggestion takes no account of the fact that a given experimental result may be generalised in infinitely many ways. This is a substantial objection since, clearly, different generalisations give rise to different expectations about the outcomes of future experiments. Musgrave's account is incomplete without some rule to specify in each case the appropriate generalisation that should be formulated and adopted. Finally, the decision to designate the generalisation background knowledge, with the consequent effect on our evaluation of other theories and on our future conduct regarding, for example, whether to repeat certain experiments, is comprehensible only if we have invested some confidence in the theory. But then Musgrave's account tacitly calls on the same kind of inductive considerations as it was designed to circumvent, so its aim is defeated.

■ e THE PROBABILITY OF THE EVIDENCE

The degree to which h is confirmed by e depends, according to Bayesian theory, on the extent to which $P(e \mid h)$ exceeds $P(e)$, that is, on how much more probable e is relative to the hypothesis and background assumptions than it is relative just to background assumptions. Another way of putting this is to say that confirmation is correlated with how much more probable the evidence is if the hypothesis is true than if it is false. This is obvious from Bayes's Theorem when it is reformulated as follows:

$$\frac{P(h \mid e)}{P(h)} = \frac{P(e \mid h)}{P(e)} = \frac{1}{P(h) + \dfrac{P(e \mid \sim h)}{P(e \mid h)} P(\sim h)} \, .$$

These facts are reflected in the everyday experience that information that is particularly unexpected or surprising unless some hypothesis is assumed to be true, supports that hypothesis with particular force. Thus, if a soothsayer predicts that you will meet a dark stranger sometime and you do in fact, your faith in his powers of precognition would not be much enhanced: you would probably continue to think his predictions were just the result of guesswork. However, if the prediction also gave the correct number of hairs on the head of that stranger, your previous scepticism would no doubt be severely shaken.

Cox (1961, p. 92) illustrated this point with an incident in *Macbeth*. The three witches, using their special brand of divination, predicted to Macbeth that he would soon become both Thane of Cawdor and King of Scotland. He finds both these prognostications almost impossible to believe:

> By Sinel's death, I know I am Thane of Glamis,
> But how of Cawdor?
> The Thane of Cawdor lives, a prosperous gentleman,
> And to be King stands not within the prospect of belief,
> No more than to be Cawdor.

But a short time later he learns that the Thane of Cawdor prospered no longer and was in fact dead and that he, Macbeth, has succeeded to the title. As a result, Macbeth's attitude to the witches' powers is entirely altered and he comes to believe in their other predictions and in their ability to foresee the future.

The following, more scientific, example was used by Jevons (1874, vol. 1, pp. 278–279) to illustrate the dependence of confirmation on the improbability of the evidence. The distinguished scientist, Charles Babbage, examined numerous logarithmic tables published over two centuries in various parts of the world. He was interested in whether they were derived from the same source or had been worked out independently. Babbage (1827) found the same six errors in all but two and drew the "irrestistible" conclusion that, apart from these two, all the tables originated in a common source.

Babbage's reasoning was interpreted by Jevons roughly as follows. The theory, t_1, which says of some pair of logarithmic tables that they had a common origin, is moderately likely, in view of the immense amount of labour needed to compile such tables ab initio, and for a number of other reasons. The alternative independence theory might take a variety of forms, each attributing different probabilities to the occurrence of errors in various positions in the table. The only one of these which seems at all likely would assign each place an equal probability of exhibiting an error and would, moreover, regard these errors as being more or less independent. Call this theory t_2 and let e^i be the evidence of i common errors in the tables. The posterior probability of t_1 is inversely proportional to $P(e^i)$, which, under the assumption of only two rival hypotheses, can be expressed as $P(e^i) = P(e^i \mid t_1)P(t_1) + P(e^i \mid t_2)P(t_2)$. (This is the theorem of total probability—*see* Chapter 2, section **e.**) Since t_1 entails e^i, $P(e^i) = P(t_1) + P(e^i \mid t_2)P(t_2)$. The quantity $P(e^i \mid t_2)$ clearly decreases with increasing i. Hence $P(e^i)$ diminishes and tends to $P(t_1)$, as i increases; and so e^i becomes increasingly powerful evidence for t_1, a result which agrees with scientific intuition.

In fact, scientists seem to regard a few shared mistakes in different mathematical tables as so strongly indicative of a common source that at least one compiler of such tables attempted to protect his copyright by deliberately incorporating three minor errors "as a trap for would-be plagiarists" (L. J. Comrie, quoted by Bowden, 1953, p. 4).

The relationship between how surprising a piece of evidence is on background assumptions and its power to confirm a hypothesis is a natural consequence of the Bayesian theory and was not deliberately built in. On the other hand, approaches that eschew probabilistic assessments of hypotheses and attempt to base scientific method on deductive logic alone seem constitutionally incapable of accounting for the phenomenon. Such approaches would need to be able, first, to discriminate between items of evidence on grounds other than their deductive or probabilistic relation to a hypothesis. And having established such a basis for discriminating, they must show a connection with confirmation. The objectivist school has more or less dodged this challenge. An exception is Popper. In tackling the problem, he moved partway towards Bayesianism; however, the concessions he made were insufficient. Thus Popper conceded that, in regard to confirmation, the significant quantities are $P(e \mid h)$ and $P(e)$, and he even measured the degree to which e confirms h (or "corroborates" it, to use Popper's preferred term) by the difference between these quantities. (Popper, 1959a, appendix *ix)

But Popper never stated explicitly what he meant by the probability of evidence. On the one hand, he would never have allowed it to have a subjective connotation, for that would have compromised the supposed objectivity of science; on the other hand, he never worked out what objective significance the term could have. His writings suggest that he had in mind some purely logical notion of probability, but as we saw in Chapter 3, there is no adequate account of logical probability. Popper also never explained satisfactorily why a hypothesis benefits from improbable evidence or, to put the objection another way, he failed to provide a foundation in non-Bayesian terms for the Bayesian confirmation function which he appropriated. (For a discussion and decisive criticism of Popper's account, see Grünbaum, 1976.)

The Bayesian position has recently been misunderstood to imply that if some evidence is known, then it cannot support any hypothesis, on the grounds that known evidence must have unit probability. That the objection is based on a misunderstanding is explained in Chapter 11, where a number of other criticisms of the Bayesian approach will be rebutted.

■ f THE RAVENS PARADOX

That evidence supports a hypothesis more the greater the ratio $\dfrac{P(e \mid h)}{P(e)}$ scotches a famous puzzle first posed by Hempel (1945) and known as the *Paradox of Confirmation* or sometimes as the *Ravens Paradox*. It was called a paradox because its premisses were regarded as extremely plausible, despite their counterintuitive, or in some versions contradictory, implications, and the reference to ravens stems from the paradigm hypothesis ('All ravens are black') which is frequently used to expound the problem. The difficulty arises from three assumptions about confirmation. They are as follows:

1. Hypotheses of the form 'All R's are B' are confirmed by the evidence of something that is both R and B. For example, 'All ravens are black' is confirmed by a black raven. (Hempel called this Nicod's condition, after the philosopher, Jean Nicod.)
2. Logically equivalent hypotheses are confirmed by the same evidence. (This is the Equivalence condition.)
3. Evidence of some object not being R does not confirm 'All R's are B'.

We shall describe an object that is both black and a raven with the term RB. Similarly, a non-black, non-raven will be denoted $\overline{RB}$. A contradiction arises for the following reasons: RB confirms 'All R's are B', on account of the Nicod condition. According to the Equivalence condition, it also confirms 'All non-B's are non-R's', since the two hypotheses are logically equivalent. But contradicting this, the third condition implies that RB does not confirm 'All non-B's are non-R's'.

The contradiction may be avoided by revoking the third condition. (We shall note later another reason for not holding on to it.) However, although the remaining conditions are compatible, they have a consequence which many philosophers have regarded as blatantly false, namely that a non-black, non-raven (say, a red herring or a white shoe) can confirm the hypothesis that all ravens are black. (The argument is this: 'All non-B's are non-R' is equivalent to 'All R's are B'; according to the Nicod condition, the first is confirmed by $\overline{RB}$; hence, by the Equivalence condition, so is the second.)

If non-black, non-ravens support the raven hypothesis, this seems to imply the paradoxical result that one could investigate that and other generalisations of a similar form *just as well* by observing white paper and red ink from the comfort of one's writing desk as by studying ravens on the wing. However, this would be a non sequitur. For the fact that RB and $\overline{RB}$ both confirm a hypothesis does not imply that they do so with equal force. Once it is recognised that confirmation is a matter of degree, the conclusion is no longer so counterintuitive, because it is compatible with $\overline{R}\overline{B}$ confirming 'All R's are B', but to a minuscule and negligible degree.

In fact, this is what Bayesians have maintained. In the particular case of the hypothesis of the ravens, Mackie (1963) argued that since non-black, non-ravens form such a numerous class compared with black ravens, it is almost (but not absolutely) certain that a random object about which we know nothing will turn out to be neither black nor a raven, but relatively unlikely that it will be a black raven. Hence, for a Bayesian, both kinds of object confirm 'All ravens are black', but non-black, non-ravens do so only minutely.

Although the Nicod and Equivalence conditions are not undermined by their implication that the raven hypothesis is confirmed by non-ravens, there are nevertheless good reasons for rejecting the Nicod condition. (The Equivalence condition seems incontestable.) As Good (1961) first demonstrated, 'All R's are B' is not necessarily confirmed by an RB and, contrary to Nicod, could even be disconfirmed by such an instance. Consider the following example of this effect, which we have taken, with some modification, from Swinburne (1971): 'All grasshoppers are located outside the county of Yorkshire'. The observation of a grasshopper just beyond the county border is an instance of this generalisation and, according to Nicod, confirms it. But it might be more reasonably argued that since there are no border controls restricting the movement of grasshoppers, the observation of one on the edge of the county increases the probability that others have actually entered, and hence undermines the hypothesis. In Bayesian terms, this is a case where the probability of some datum is reduced by a hypothesis (that is, $P(e \mid h) < P(e)$) which is therefore disconfirmed (in other words, $P(h \mid e) < P(h)$).

The grasshopper example also provides an instance where a datum of the type $\overline{R}B$ confirms a generalisation of the form 'All R's are B'. Imagine that an object which looks for all the world like a grasshopper were found hopping about just outside Yorkshire and that it turned out to be some other sort of insect. The discovery that the object was not a grasshopper would be relatively unlikely unless the grasshopper hypothesis were true (hence, $P(e) < P(e \mid h)$); thus it would confirm that hypothesis. If the deceptively grasshopper-like object were within the county boundary, the same conclusion would follow, though the degree of confirmation would be greater. This shows that 'All R's are B' may also be confirmed by a datum of the type $\overline{R}\overline{B}$. Hence, the impression that non-R's never confirm such hypotheses may be dispelled.

It is sometimes maintained that 'All ravens are black' would be differently confirmed if a known raven were revealed to be black than if an object were first observed to be black and later found to be a raven. For instance, Horwich, who denoted the first datum $R*B$ and the second $RB*$, argued that the former is the more powerfully confirming instance, on the alleged grounds that only it subjects the hypothesis to the risk of falsification, for the raven could have turned out to be non-black, in which case the hypothesis would have been refuted. By contrast, Horwich said, the latter does not jeopardise the hypothesis, for the black object is compatible with the hypothesis whether it is a raven or not.

This argument, however, is specious. The observation of an object that enquiry reveals to be a black raven poses absolutely no risk of refutation to the hypothesis, however the enquiry was conducted. The only difference between $R*B$ and $RB*$ is in the point at which one learns that the hypothesis has not been refuted. This does not seem to us a sufficient reason to distinguish the two data from the point of view of their confirming power. To do so would appear to depart from normal practice, for scientists do not as a rule attach any importance to the distinction. (For a fuller discussion of this point, the reader is referred to Chapter 11, section **g**.)

Our conclusions are, first, that the supposedly paradoxical consequences of Nicod's condition and the Equivalence condition are not problematic, and, secondly, that there are separate reasons for rejecting Nicod's condition, which, moreover, conform to Bayesian principles.

■ g THE DESIGN OF EXPERIMENTS

Not every experiment is equally worth doing and because of the expense that experiments often necessitate, both in labour and in equipment, careful attention is frequently devoted to their design, in order to ensure that they will yield information economically.

What is a well-designed experiment? The natural answer is that it is an experiment which stands a good chance of producing a decisive or, at least, an almost decisive result. The experiment should be decisive in the sense that one hypothesis becomes certainly true or at least almost certainly true, or that as many as possible of the initially most plausible hypotheses become (almost) certainly not true. This allows for the possibility that a poorly constructed experiment may, unexpectedly, produce decisive evidence, while a well-designed experiment may yield an outcome which is quite indecisive.

The considerations that are pertinent to the design of efficient experiments can be appreciated by referring to Bayes's Theorem. Suppose n rival hypotheses, $h_1, \ldots, h_n$, are being entertained and that these are regarded as the only serious contenders for the truth, in the sense that their total probability is 1 or close to 1. We are, as we said, interested in acquiring decisive evidence, that is, the kind of evidence, e, which makes $P(h_i \mid e)$ approach 1 for some h_i or which brings as many as possible of the terms $P(h_j \mid e)$ close to 0. Consider now an experiment, one of whose possible outcomes, e, would have the effect of massively confirming or disconfirming one of the hypotheses. Such evidence would be decisive in our sense. Clearly, the larger $P(e)$, the greater the probability of achieving a decisive result and, hence, the better the experiment.

However, a slightly odd fact emerges at this point. In order to confirm a hypothesis strongly, one requires evidence e for which $P(e)$ is low, relative to $P(e \mid h)$. On the other hand, in order for the experiment to be worth doing at all, $P(e)$ should be moderately high. Therefore, two separate considerations determine how well designed an experiment is, and these frequently pull in opposite directions.

When deciding which experiment to perform, one must also take at least three other factors into account: the cost of the experiment; the morality of carrying it out; and the value, both theoretical and practical, of the hypotheses one is interested in. Bayes's Theorem, of course, implies nothing about how these separate factors should be balanced.

■ h THE DUHEM PROBLEM

h.1 The Problem

The so-called Duhem (or Duhem-Quine) problem is a problem for theories of science of the type associated with Popper, which emphasise the power of certain evidence to refute a hypothesis. According to Popper's influential views, the characteristic of a theory which makes it 'scientific' is its falsifiability: "Statements or systems of statements, in order to be ranked as scientific, must be capable of conflicting with possible, or conceivable, observations" (Popper, 1963, p. 39). And, claiming to apply this criterion, Popper (1963, ch. 1) judged Einstein's gravitational theory to be scientific and Freud's psychology, unscientific.

There is a strong flavour of commendation about the term scientific which has proved extremely misleading. For a theory which is scientific in Popper's sense is not necessarily true, or even probably true or so much as close to the truth, nor can it be said definitely that it is likely to lead to the truth. In fact, there seems to be no conceptual connection between a theory's capacity to pass Popper's test of scientificness and its having any epistemic or inductive value. There is little alternative, then, so far as we can see, to regarding Popper's demarcation between scientific and unscientific statements as part of a theory about the content and character of what is usually termed science, not as having any normative significance.

Yet as a contribution to understanding the methods of science, Popper's ideas bear little fruit. His central claim was that scientific theories are falsifiable by "possible, or conceivable, observations". This poses a difficulty, for an observation can only falsify a theory (that is, conclusively demonstrate its falsity) if it is itself conclusively certain. But observations cannot be conclusively certain. For instance, the statement 'The hand on this dial is pointing to the numeral 6' is clearly fallible—it is unlikely, but possible, that the person reporting it missaw the position of the hand. The same is true of introspective perceptual reports, such as 'In my visual field there is now a silvery crescent against a dark blue background'. It has recently been maintained (Watkins, 1984, pp. 79 and 248) that this and similar statements "may rightly be regarded by their authors when they make them as infallibly true". But this is not so, for it is possible, though not probable, that the introspector has misremembered and mistaken the shape he usually describes as a crescent or the sensation he usually receives on reporting a blue image. These and other sources of error ensure that introspective reports are not exempt from the rule that non-analytic statements are fallible.

Of course, the kinds of observation statement we have mentioned, if asserted under appropriate circumstances, would never be seriously doubted. That is, although they could be false, they have a force and immediacy that carries conviction; they are 'morally certain', to use the traditional phrase. But if observation statements are merely indubitable, then whether a theory is regarded as refuted by observational data or not rests ultimately on a subjective feeling of certainty. The fact that such convictions are so strong and uncontroversial may disguise their fallibility, but cannot undo it. Hence, no theory is falsifiable, for none could be conclusively shown to be false by empirical observations. In practice the closest one could get to a refutation would be arriving at the conclusion that a theory that clashes with almost certainly true observations is almost certainly false.

A second objection to Popper's falsifiability criterion, and the one upon which we shall focus for its more general interest, is that it describes as unscientific most of those theories which are usually deemed science's greatest achievements. This is the chief aspect of the well-known criticisms advanced by Polanyi, Kuhn, and Lakatos, amongst others. They have pointed out that, as had already been established by Duhem (1905), many notable theories of science are not falsifiable by what would generally be regarded as observation statements, even if those statements were infallibly true. Predictions drawn from Newton's laws or from the kinetic theory of gases turn out to depend not only on those theories but also on certain auxiliary theories. Hence, if such predictions fail, one is not compelled by logic to infer that the main theory is false, for the fault may lie with one or more of the auxiliary assumptions. The history of science has many occasions when an important theory led to a false prediction and where that theory, nevertheless, was not blamed for the failure. In such cases we find that one or more of the auxiliary assumptions used to derive the prediction was taken to be the culprit. The problem that arose from Duhem's investigations was which of the several distinct theories involved in deriving a false prediction should be regarded as the false element or elements in the assumptions.

h.2 Lakatos's and Kuhn's Treatment of the Duhem Problem

Lakatos examined in detail the way that scientists react to

anomalies; indeed, he made it a central feature of what he referred to as his "methodology of scientific research programmes". Lakatos claimed that scientific research of the most significant kind usually proceeds in what he called "research programmes". A research programme takes the form of a central or "hard core" theory, together with an associated "protective belt" of auxiliary assumptions. The function of the latter is to combine with the hard core in order to draw out specific predictions, which can then be checked by experiment. The auxiliary assumptions are described as protective because during a research programme's lifetime they, not the central theory, are revised if a prediction is shown to be false.

Lakatos suggested Newtonian physics as an example of a research programme, the three laws of mechanics and the law of gravitation constituting the hard core, while various optical theories, assumptions about the number and positions of the planets, and so forth, he included in the protective belt. He also described a set of heuristic rules by which the research programme dealt with anomalies and advanced into new areas.

Kuhn's famous theory of scientific paradigms is similar to the methodology we have just described and was probably its inspiration. Both Lakatos and Kuhn were impressed that scientists tend to give the benefit of the doubt to some, especially fundamental, theories when these encounter anomalies—and both argued that such theories exerted a commanding influence over whole areas of scientific research. Lakatos's methodology has the advantage in that it describes scientific research programmes in some detail and that it analyses their modes of action; whereas Kuhn left his corresponding notion of a paradigm somewhat vague in comparison.

Lakatos also outlined criteria of success for a research programme. He held that it was perfectly legitimate to systematically treat the hard core as the innocent party in a refutation, provided the research programme occasionally leads to successful novel predictions or to successful or "non–ad hoc" explanations of existing data. Lakatos called such programmes "progressive".

The sophisticated falsificationist [which Lakatos counted himself as] ... sees nothing wrong with a group of brilliant scientists conspiring to pack everything they can into their favourite research programme ('conceptual framework', if you wish) with a sacred hard core. As long as their genius—and luck—en-

> ables them to expand their programme *'progressively'*, while
> sticking to its hard core, they are allowed to do it. (Lakatos,
> 1970, p. 187)

If, on the other hand, the programme persistently produced
false predictions, or if its explanations were habitually ad hoc,
Lakatos called it "degenerating". (We shall devote the next
section to the notion of ad hocness.) Lakatos employed these
tendentious terms even though he never succeeded in substan-
tiating their intimations of approval and disapproval, and in
the end he seems to have abandoned the attempt and settled
on the more modest claim that, as a matter of historical fact,
progressive programmes have usually been well regarded by
scientists, while degenerating ones were distrusted and eventually dropped.

This last claim has, it seems to us, some truth to it, as
evidenced, for example, by the case studies in the history of
science included in Howson (1976). But although Lakatos and
Kuhn identified and described an important aspect of scientific
work, they provided no rationale or explanation for it. For instance, Lakatos was never able to explain why a research programme's occasional predictive or explanatory success could
compensate for numerous failures, nor could he specify how
many such successes are needed to convert a degenerating programme into a progressive one. (They should occur "now and
then", he said.) Hence, although the methodology of scientific
research programmes points to some of the factors relevant to
scientific change, it provides no explanation.

Lakatos was also unable to explain why some theories are
raised to the status of the hard core of a research programme
and are defended by a protective belt of hypotheses, while
others are left to their own devices. From Lakatos's writings,
one could think that the question is decided by the scientist's
mere whim (Lakatos called it a "methodological fiat"). Unfortunately, this suggests that it is a perfectly canonical scientific
practice to set up any theory whatever as the hard core of a
research programme, or as the central pattern of a paradigm,
and to blame all empirical difficulties on auxiliary theories.
This is far from being the case.

h.3 The Duhem Problem Solved by Bayesian Means

The questions left unanswered by Lakatos are answered with
the help of Bayes's Theorem, as Dorling (1979) has shown. First
we shall consider how the probabilities of several theories are
altered when, as a group, they have been refuted.

Suppose a theory, t, and an auxiliary hypothesis, a, together imply an empirical consequence, which is shown to be false by the observation of the outcome e. Let us assume that while the combination t & a is refuted by e, the two components taken separately are not refuted. We wish to consider the separate effects wrought on the probabilities of t and a by the adverse evidence e. The comparisons of interest here are between $P(t \mid e)$ and $P(t)$ and between $P(a \mid e)$ and $P(a)$. The conditional probabilities can be expressed using Bayes's Theorem, as follows:

$$P(t \mid e) = \frac{P(e \mid t)P(t)}{P(e)} \qquad P(a \mid e) = \frac{P(e \mid a)P(a)}{P(e)}.$$

In order to evaluate the posterior probabilities of t and of a, one must first determine the values of the various terms on the right-hand sides of these equations. Before doing this, it is worth noting that these expressions convey no expectation that the refutation of t & a jointly considered will in general have a symmetrical effect on the separate probabilities of t and of a, nor any reason why the degree of asymmetry may not be very large in some cases. Also, the expressions allow one to discern the factors that determine which hypothesis suffers most in the refutation. In particular, the probability of t changes very little if $P(e \mid t) \approx P(e)$, while that of a is reduced substantially just in case $P(e \mid a)$ is substantially less than $P(e)$.

A historical example might best illustrate how a theory that produces a false prediction may still remain very probable; we shall, in fact, use an example that Lakatos (1970, pp. 138–140, and 1968, pp. 174–175) drew heavily on. In 1815, William Prout, a medical practitioner and chemist, advanced the hypothesis that the atomic weights of all the elements are whole-number multiples of the atomic weight of hydrogen, the underlying assumption being that all matter is built out of different combinations of some basic element. Prout believed hydrogen to be that fundamental building-block, though the idea was entertained by others that a more basic element might exist, out of which hydrogen itself was composed. Now the atomic weights recorded at the time, though close to being integers when expressed as multiples of the atomic weight of hydrogen, did not match Prout's hypothesis exactly. However, these deviations from a perfect fit failed to convince Prout that his hypothesis was wrong; he instead took the view that there were faults in the methods that had been used to measure the relative weights of atoms. Thomas Thomson drew a similar conclusion. Indeed, both he and Prout went so far as to adjust several reported atomic weights in order to bring them into line with Prout's hypothesis. For instance, instead of accepting 0.829 as the atomic weight (expressed as a proportion of the weight of an atom of oxygen) of the element boron, which was the experimentally reported value, Thomson (1818, p. 340) preferred 0.875 "because it is a multiple of 0.125, which all the atoms seem to be". (Thomson erroneously took 0.125 as the atomic weight of hydrogen, relative to that of oxygen.) Similarly, Prout adjusted the measured atomic weight of chlorine, which (relative to hydrogen) was 35.83, to 36.

Thomson's and Prout's reasoning can be explained as follows: Prout's hypothesis t, together with an appropriate assumption a asserting the accuracy (within specified limits) of the measuring technique, the purity of the chemicals employed, and so forth, implies that the measured atomic weight of chlorine (relative to hydrogen) is a whole number. Suppose, as was the case in 1815, that chlorine's measured atomic weight was 35.83, and call this the evidence e. It seems that chemists of the early nineteenth century, such as Prout and Thomson, were fairly certain about the truth of t, but less so of a, though more sure that a is true than that it is false. Contemporary near-certainty about the truth of Prout's hypothesis is witnessed by the chemist J. S. Stas. He reported (1860, p. 42) that "In England the hypothesis of Dr Prout was almost universally accepted as absolute truth", and he confessed that when he started researching into the matter, he himself had "had an almost absolute confidence in the exactness of Prout's principle" (1860, p. 44). (Stas's confidence eventually faded after many years' experimental study, and by 1860 he had "reached the complete conviction, the entire certainty, as far as certainty can be attained on such a subject that Prout's law . . . is nothing but an illusion", 1860, p. 45.) It is less easy to ascertain how confident Prout and his contemporaries were in the methods by which atomic weights were measured, but it is unlikely that this confidence was very great, in view of the many clear sources of error and the failure of independent measurements generally to produce identical results. On the other hand, chemists of the time must have felt that their methods for determining atomic weights were more likely to be accurate than not, otherwise they would not have used them. For these reasons, we conjecture that $P(a)$ was of the order of 0.6 and that $P(t)$ was around 0.9, and these are the figures we shall work with. It should be stressed that these numbers and those we shall assign to other probabilities are intended chiefly to illustrate how Bayes's Theorem resolves Duhem's problem; nevertheless, we believe them to be sufficiently accurate to throw light on the progress of Prout's hypothesis. As we shall see, the results we obtain are not very sensitive to variations in the assumed prior probabilities.

In order to evaluate the posterior probabilities of t and of a, one must fix the values of the terms $P(e \mid t)$, $P(e \mid a)$ and $P(e)$. These can be expressed, using the Theorem on Total Probability (Chapter 2, section **e**), as follows:

$$P(e) = P(e \mid t)P(t) + P(e \mid \sim t)P(\sim t)$$

$$P(e \mid t) = P(e \& a \mid t) + P(e \& \sim a \mid t)$$

$$= P(e \mid t \& a)P(a \mid t) + P(e \mid t \& \sim a)P(\sim a \mid t)$$

$$= P(e \mid t \& a)P(a) + P(e \mid t \& \sim a)P(\sim a)$$

Since $t \& a$, in combination, is refuted by e, the term $P(e \mid t \& a)$ is zero. Hence:

$$P(e \mid t) = P(e \mid t \& \sim a)P(\sim a).$$

It should be noted that in deriving the last equation but one, we have followed Dorling in assuming that t and a are independent, that is, that $P(a \mid t) = P(a)$ and, hence, $P(\sim a \mid t) = P(\sim a)$. This seems to accord with many historical cases and is clearly right in the present case. By parallel reasoning to that employed above, we may derive the results:

$$P(e \mid a) = P(e \mid \sim t \& a)P(\sim t)$$

$$P(e \mid \sim t) = P(e \mid \sim t \& a)P(a) + P(e \mid \sim t \& \sim a)P(\sim a).$$

Provided the following terms are fixed, which we have done in a tentative way, to be justified presently, the posterior probabilities of t and of a can be determined:

$$P(e \mid \sim t \& a) = 0.01$$

$$P(e \mid \sim t \& \sim a) = 0.01$$

$$P(e \mid t \& \sim a) = 0.02.$$

The first of these gives the probability of the evidence if Prout's hypothesis is not true but if the method of atomic weight measurement is accurate. Such probabilities were explicitly considered by some nineteenth century chemists, and they typically took a theory of random assignment of atomic weights as the alternative to Prout's hypothesis (e.g., Mallet, 1880); we shall follow this. Suppose it had been established for certain that the atomic weight of chlorine lay between 35 and 36. (The final results we obtain respecting the posterior probabilities of t and a are, incidentally, not affected by the width of this interval.) The random-allocation theory would assign equal probabilities to the atomic weight of an element lying in any 0.01-wide interval. Hence, on the assumption that a is true, but t false, the probability that the atomic weight of chlorine lies in the interval 35.825 to 35.835 is 0.01. We have assigned the same value to $P(e \mid \sim t \ \& \ \sim a)$ on the grounds that if a were false because, say, some of the chemicals were impure or the measuring techniques faulty, then, still assuming t to be false, one would not expect atomic weights to be biased towards any particular part of the interval between adjacent integers.

We have set the probability $P(e \mid t \ \& \ \sim a)$ rather higher, at 0.02. The reason for this is that although some impurities in the chemicals and some degree of inaccuracy in the method of measurement were moderately likely in the early nineteenth century, chemists certainly would not have considered their techniques entirely haphazard. Thus if Prout's hypothesis were true, but the measuring technique imperfect, the measured atomic weights would have been likely to deviate somewhat from integral values; but the greater the deviation, the less likely, on these assumptions, so the probability of an atomic weight lying in any part of the 35–36 interval would not be distributed uniformly over the interval, but would be more concentrated around the whole numbers. Let us proceed with the figures we have assumed for the crucial probabilities.* We thus obtain:

$$P(e \mid \sim t) = \quad 0.01 \times 0.6 + 0.01 \times 0.4 \quad = 0.01$$

$$P(e \mid t) = \quad 0.02 \times 0.4 \quad = 0.008$$

$$P(e \mid a) = \quad 0.01 \times 0.1 \quad = 0.001$$

$$P(e) = \quad 0.008 \times 0.9 + 0.01 \times 0.1 \quad = 0.0082$$

---

*As a matter of fact, it is not the particular values taken by the three probability terms that are important, but their *relative* values. Thus we would arrive at the same posterior probabilities for a and t with the weaker assumptions that $P(e \mid \sim t \ \& \ a) = P(e \mid \sim t \ \& \ \sim a) = \frac{1}{2} \, P(e \mid t \ \& \ \sim a)$.

Finally, Bayes's Theorem enables us to derive the posterior probabilities in which we were interested:

$P(t \mid e) = 0.878$ (Recall that $P(t) = 0.9$)

$P(a \mid e) = 0.073$ (Recall that $P(a) = 0.6$)

These striking results show that evidence of the kind we have described may have a sharply asymmetric effect on the probabilities of t and of a. The initial probabilities we assumed seem appropriate for chemists such as Prout and Thomson, and if they are correct, the results deduced from Bayes's Theorem explain why those chemists regarded Prout's hypothesis as being more or less undisturbed when certain atomic-weight measurements diverged from integral values, and why they felt entitled to adjust those measurements to the nearest whole number. Fortunately, these results are relatively insensitive to changes in our assumptions, so their accuracy is not a vital matter as far as our explanation is concerned. For example, if one took the initial probability of Prout's hypothesis (t) to be 0.7, instead of 0.9, keeping the other assignments, we find that $P(t \mid e) = 0.65$, while $P(a \mid e) = 0.21$. Thus, as before, after the refutation, Prout's hypothesis is still more likely to be true than false, and the auxiliary assumptions are still much more likely to be false than true. Other substantial variations in the initial probabilities produce similar results, though with so many factors at work, it is difficult to state concisely the conditions upon which these results depend without just pointing to the equations above. Thus Bayes's Theorem provides a model to account for the kind of scientific reasoning that gave rise to the Duhem problem. And the example of Prout's hypothesis, as well as others that Dorling (1979 and 1982) has described, show, in our view, that the Bayesian model is essentially correct. By contrast, non-probabilistic theories seem to lack entirely the resources that could deal with Duhem's problem.

A fact that emerges when slightly different values are assumed for the various probabilities in the Prout's hypothesis example is that one or other of the theories may actually become more probable after the conjunction t & a has been refuted. For instance, when $P(e \mid t$ & $\sim a)$ equals 0.05, the other probabilities being assigned the same values as before, the posterior probability of t is 0.91, which exceeds its prior probability. This may seem bizarre but, as Dorling (1982) has argued, it is not so odd when one bears in mind that the refuting evidence normally contains a good deal more information than is required merely to disprove t & a and that this extra information may be confirmatory. In general, such confirmation occurs when $P(e) < P(e \mid t)$, which is easily shown to be equivalent to the condition $P(e \mid t) > P(e \mid \sim t)$. In other words, when evidence is easier to explain (in the sense of having a higher probability) if a given hypothesis is true than if it is not, then that theory is confirmed by the evidence.

■ I GOOD DATA, BAD DATA, AND DATA TOO GOOD TO BE TRUE

Good data. The marginal influence which we have seen an anomalous observation may exert on the probability of a theory is to be contrasted with the dramatic effect that a confirmation can have. For instance, if the measured atomic weight of chlorine had been a whole number, in line with Prout's hypothesis, so that now $P(e \mid t \ \& \ a)$ is one instead of zero, and if the probabilities we assigned were kept, the probability of the hypothesis would have shot up from a prior of 0.9 to 0.998. And, even more dramatically, if the prior probability of t had been 0.7, its posterior probability would have risen to 0.99. The existence of this asymmetry between anomalous and confirming instances was highlighted with particular vigour by Lakatos, who regarded it as being of the greatest significance in science and as one of the characteristic features of a research programme; Lakatos maintained that a scientist involved in such a programme typically "forges ahead with almost complete disregard of 'refutations'", provided he is occasionally rewarded with successful predictions (1970, p. 137): he is "encouraged by Nature's YES, but not discouraged by its NO" (1970, p. 135). As we have indicated, we believe there to be much truth in Lakatos's observations; however, they are merely incorporated without explanation into his methodology, while the Bayesian has a simple and plausible explanatory model.

Bad data. An interesting fact that emerges from the Bayesian analysis is that a successful prediction derived from a combination of two theories, say t and a, does not always redound to the credit of t, even if the prior probability of the evidence is small; indeed, it can even undermine it. We may illustrate this by referring again to the example of Prout's hypothesis.

Suppose the atomic weight of chlorine were 'measured', not in the old-fashioned chemical way, but by concentrating hard on the element in question and picking a number in some random fashion from a given range of numbers. And let us assume that this method assigns a whole-number value to the atomic weight of chlorine. This is just what one would predict on the basis of Prout's hypothesis, if the outlandish measuring technique were reliable. But reliability is obviously most unlikely and it is equally obvious that, as a result, the measured atomic weight of chlorine adds practically nothing to the probability of Prout's hypothesis, notwithstanding its integral value. This intuition is upheld by Bayes's Theorem, as a simple calculation based on the above formulas shows. (As before, let t be Prout's hypothesis and a the assumption that the measuring technique is accurate. Then set $P(e \mid t \ \& \ {\sim}a) = P(e \mid {\sim}t \ \& \ {\sim}a) = P(e \mid {\sim}t \ \& \ a) = 0.01$, for reasons similar to those stated earlier, and let $P(a)$ be very small, say 0.0001, for obvious reasons. It then follows that $P(t)$ and $P(t \mid e)$ are equal to two decimal places.) This example shows that Leibniz was wrong to declare as a general principle that "It is the greatest commendation of an hypothesis (next to truth) if by its help predictions can be made even about phenomena or experiments not tried". Leibniz and Lakatos, who quoted these words with approval (1970, p. 123), seem to have overlooked the fact that if a prediction can be deduced from a hypothesis only with the assistance of highly questionable auxiliary claims, then that hypothesis often accrues very little credit. This explains why the various sensational predictions which Velikovsky drew from his theory of planetary collisions failed to impress most serious scholars, even when some of those predictions were to their amazement fulfilled. For instance, Velikovsky's prediction of the existence of large quantities of petroleum on the planet Venus relied not only on his pet theory that various natural disasters in the past had been caused by collisions between the earth and a comet, but also on a number of unsupported and not very plausible assumptions, such as that the comet in question originally carried hydrogen and carbon, that these had been converted to petroleum by electrical discharges supposedly created in the violent impact with the earth, that the comet had later evolved into the planet Venus, and some others (Velikovsky, 1950, p. 351). (More details of Velikovsky's theory are given in the next section.)

Data too good to be true. Data are sometimes said to be 'too good to be true' when they seem to fit a favoured hypothesis more perfectly than it is reasonable to expect. For instance, suppose all the atomic weights listed in Prout's paper had been whole numbers, exactly. Such a result almost looks as if it was designed to impress, and it is just for this reason that it fails to. We may analyse this response as follows. Let e be the evidence of, say, 20 atomic-weight measurements, each a perfect whole number. No one could have regarded precise atomic weights measured at the time as absolutely reliable. The most natural view would have been that such measurements are subject to experimental error and, hence, that they would give a certain spread of results about the true value. On this assumption, which we shall label a', it is extremely unlikely that numerous independent atomic-weight measurements would all produce whole numbers, even if Prout's hypothesis were true. So $P(e \mid t \& a')$ is extremely small and, clearly, $P(e \mid {\sim}t \& a')$ would be no larger. Now a' has many possible alternatives, one of the more plausible (though initially it might not be very plausible) being that the experiments were consciously or unconsciously rigged in favour of Prout's hypothesis. If this were the only significant alternative (and so, in effect, equivalent to ${\sim}a'$), $P(e \mid t \& {\sim}a')$ would be very high, as would $P(e \mid {\sim}t \& {\sim}a')$. It follows from the equations on pages 99–100 above that

$$P(e \mid t) \approx P(e \mid t \& {\sim}a')P({\sim}a') \text{ and}$$
$$P(e \mid {\sim}t) \approx P(e \mid {\sim}t \& {\sim}a')P({\sim}a')$$

and, hence,

$$P(e) \approx P(e \mid t \& {\sim}a')P({\sim}a')P(t) +$$
$$P(e \mid {\sim}t \& {\sim}a')P({\sim}a')P({\sim}t).$$

Now, presumably the rigging of the results to produce whole numbers, if it took place, would produce whole numbers equally effectively whether t was true or not; in other words,

$$P(e \mid t \& {\sim}a') = P(e \mid {\sim}t \& {\sim}a');$$

hence

$$P(e) \approx P(e \mid t \& {\sim}a')P({\sim}a').$$

Therefore,

$$P(t \mid e) = \frac{P(e \mid t)P(t)}{P(e)} \approx \frac{P(e \mid t \& {\sim}a')P({\sim}a')P(t)}{P(e \mid t \& {\sim}a')P({\sim}a')} = P(t)$$

Thus e does not confirm t significantly, even though, in a misleading sense, it fits the theory perfectly. This is why it is said to be too good to be true. A similar calculation shows that the probability of a' is diminished and, on the assumptions that we made, this implies that the probability of the experiments having been fabricated is enhanced. (The above analysis is essentially the same as given in Dorling, 1982).

A famous case of data that were criticized for being too good to be true is that of Mendel's plant-breeding results. Mendel's genetic theory of inheritance allows one to calculate the probabilities with which certain plants would produce specific kinds of offspring. For instance, under certain circumstances, pea plants of a particular strain may be calculated to yield round and wrinkled seeds with probabilities 0.75 and 0.25, respectively. Mendel obtained seed-frequencies that matched the corresponding probabilities in this and in similar cases remarkably well, suggesting (misleadingly Fisher contended) substantial support for the genetic theory. Fisher did not believe that Mendel had deliberately falsified his results to appear in better accord with his theory than they really were. To do so, Fisher claimed, would "contravene the weight of the evidence supplied in detail by . . . [Mendel's] paper as a whole" (1936, p. 132). But Fisher thought it a "possibility among others that Mendel was deceived by some assistant who knew too well what was expected" (1936, p. 132), an explanation he backed up with some (rather meagre) independent evidence.

The argument put forward earlier to show that too-exactly whole-number atomic-weight measurements would not have supported Prout's hypothesis depends on the existence of some sufficiently plausible alternative hypothesis that explains the data better. We believe that, in general, data are too good to be true relative to one hypothesis only if there are such alternatives. This principle accords with intuition; for if the technique for eliciting atomic weights had long been established as precise and accurate, and if careful precautions had been taken against experimenter bias, all the natural alternatives to Prout's hypothesis could be discounted and the data would no longer seem suspiciously good; they would be straightforwardly good. Fisher, however, did not subscribe to the principle, at least, not explicitly; he believed that Mendel's results told against the genetic theory whatever alternative explanations might suggest themselves. Nevertheless, as just indicated, the consideration of such alternatives played a part in his argument. We shall refer again to Fisher's case against Mendel in the next chapter.

■ J AD HOC HYPOTHESES

As we have seen, an important scientific theory which, in combination with other assumptions, has made a false prediction may nevertheless emerge relatively unscathed, while the auxiliary hypotheses are largely discredited. (We are using such expressions in the normal way to describe how hypotheses are received, regarding them as harmless metaphors for obvious and more or less precise probabilistic notions. Thus, a hypothesis that is unscathed by negative evidence is one whose posterior and prior probabilities are similar. On the other hand, it is difficult to understand what opponents of the Bayesian approach could have in mind when they talk of theories being 'accepted' or 'retained', or 'put forward' or 'saved' or 'vindicated'.) When a set of auxiliary assumptions is discredited in a test, scientists frequently think up new assumptions which assist the main theory to explain the previously anomalous data. Sometimes these new assumptions give the impression that their role is simply to 'patch up' the theory, and in such cases Francis Bacon called them "frivolous distinctions" (1620, Book I, aphorism xxv). More recently they have been tagged 'ad hoc hypotheses', presumably because they would not have been introduced if the need to bring theory and evidence into line had not arisen. However, the term is pejorative, and hypotheses falling into the ad hoc category are very often dismissed as more or less worthless.

But although particular ad hoc theories are fairly easy to evaluate intuitively, there is controversy over what general criteria apply. Indeed, there is not even a universally accepted definition of 'ad hoc' as that term is applied to hypotheses. We shall see that the Bayesian approach clarifies the question. First let us consider a few uncontroversial examples and then deal with some general accounts of ad hocness.

j.1 Some Examples of Ad Hoc Hypotheses

Velikovsky's theory of collective amnesia. Immanuel Velikovsky, in a daring book called *Worlds in Collision* that attracted a great deal of attention some years ago, put forward the theory that the world has been subject, at various stages in its history, to cosmic disasters produced by near collisions with massive comets. One of these comets, which went on to make a distinguished career as the planet Venus, is supposed to have passed close by the earth during the Israelites' captivity in Egypt and to have caused the various remarkable events of the time, such as the ten plagues and the parting of the Red Sea. One of the theory's predictions, apparently, is that every group of people in the world will have noticed these tremendous goings-on and if they kept records at all, they would have recorded them. However, many communities failed to note in their writings anything out of the ordinary at that time, and Velikovsky, remaining convinced by his main theory, put this exceptional behaviour down to what he called a "collective amnesia". He argued that the cataclysms were so terrifying that whole peoples behaved "as if [they had] obliterated impressions that should be unforgettable". There was a need, Velikovsky said, to "uncover the vestiges" of these events, "a task not unlike that of overcoming amnesia in a single person" (1950, p. 288). Individual amnesia is the issue in the next example.

Dianetics. Dianetics is a theory that purports to analyse the causes of insanity and mental stress, which it sees as the 'misfiling' of information in inappropriate locations in the brain. By refiling these 'engrams', it claims, sanity may be restored, composure enhanced, and, incidentally, the memory vastly improved. Not surprisingly, the therapy is long and expensive, and few people have been through it and borne out the theory's claims. One triumphant success, a young student, was, however, announced by the inventor of Dianetics, L. Ron Hubbard, and in 1950 he exhibited this person to a large audience, claiming that she had a "full and perfect recall of every moment of her life". However, questions from the floor ("What did you have for breakfast on October 3, 1942?"; "What colour is Mr Hubbard's tie?", and the like) soon demonstrated that the hapless girl had a most imperfect memory. Hubbard accounted for this to what remained of the assembly by saying that when the girl first appeared on the stage and was asked to come forward "now", the word "now" had frozen her in "present time" and paralysed her ability to recall the past. (An account of the incident and of the history of Dianetics is given by Miller, 1987.)

An example from psychology. Investigations into distributions of IQ show that different groups of people vary in their average levels of measured intelligence. A number of so-called environmentalists put a low score down primarily to poor social and educational conditions. However, this explanation ran into trouble when it was discovered that a large group of Eskimos, leading a feckless, poor, and drunken existence, scored very highly on IQ tests. The distinguished biologist Peter Medawar (1974), in an effort to deflect the difficulty away from the environmentalist thesis, explained the observation by saying that an "upbringing in an igloo gives just the right degree of cosiness, security and mutual contact to conduce to a good performance in intelligence tests."

In each of these examples, the theory which replaced the refuted one seems rather unsatisfactory. It is not likely that they would have been put forward except in response to a particular empirical anomaly, and this explains the label "ad hoc", which suggests that the theory was advanced for the specific purpose of evading a difficulty. However, some theories of this kind cannot be condemned so readily. For instance, an ad hoc alteration which rescued Newtonian theory from a difficulty led directly to the discovery of a new planet and was generally deemed a great success.

The discovery of the planet Neptune. Newtonians tried unsuccessfully to account for the motion of the planet Uranus, but the difference between theory and observation exceeded the admissible limits of experimental error. Two astronomers, Adams and Leverrier, working independently, put forward a new theory which postulated the existence of a previously unthought-of planet and hence of a new source of gravitational attraction to act on Uranus. This theory was later vindicated by careful telescopic observations and studies of old astronomical maps, which revealed the presence of a planet with the anticipated characteristics. The planet was later called Neptune. (The fascinating story of this episode is told by W. M. Smart, 1947.)

j.2 A Standard Account of Ad Hocness

The salient features of the examples we are considering are that a theory t, which we can call the main theory, was combined with an auxiliary hypothesis, a, to predict e, when in fact

e' occurred, e' being incompatible with e. And in order to retain the main theory in its desired explanatory role, a new auxiliary, a', was proposed which, with t, implies e'.

Two criteria of acceptability are often applied by philosophers in such circumstances. The first is that t & a' should have test implications that are independent of the evidence that refuted t & a. The second criterion is that some of these test implications should be verified. Lakatos (1970, p. 175) called theories that failed the first, ad hoc$_1$, and those that did not satisfy the second, ad hoc$_2$. Some philosophers maintain that a theory is acceptable only if it is non-ad hoc in both of these senses (for example, Popper, 1963, pp. 244–248), while others emphasise only the first sense (for example, Hempel, 1966, p. 29). Our criticisms of this approach will not need to distinguish between the two points of view.

The term ad hoc to describe hypotheses that do not meet one or other of these conditions seems not to be an old one; its earliest occurrence in English that we know of was in 1936, in a critical review of a book of psychology. The reviewer, W. J. H. Sprott, observed that

> There is a suspicion of '*ad-hoc*-ness' about the 'explanations' [of a certain aspect of childish behaviour]. The whole point is that such an account cannot be satisfactory until we can predict the child's movements from a knowledge of the tensions, vectors and valences which are operative, *independent of our knowledge of how the child actually behaved.* So far we seem reduced to inventing valences, vectors and tensions from a knowledge of the child's behaviour. (Sprott, 1936, p. 249; our emphasis)

But although the term ad hoc is relatively new, the idea goes back at least to Bacon, who criticized as a "frivolous distinction" the type of hypothesis that is "framed to the measure of those particulars only from which it is derived". Bacon argued that a hypothesis ought to be "larger and wider" than the observations that gave rise to it and, moreover, that it should lead to new particulars. According to this criterion, the first three examples above seem to be unsatisfactory scientific developments, while the fourth does not, since the new-planet theory was supported by evidence different from that which led to the original refutation. According to this criterion, the modification which Velikovsky brought to his theory would be acceptable only if it were supported, for example, by contemporary records of amnesia or by evidence of peculiar features in the environment which we have reason to think are conducive to mass forgetfulness. Medawar's and Hubbard's theories are rather vague and seem unsusceptible of any independent test, though one must acknowledge that a closer study of those theories could reveal potential tests.

J.3 A Bayesian Account of Ad Hocness

We shall argue in the next subsection that the above account misses the characteristic of ad hoc hypotheses that determines whether they are well regarded or not by scientists. The quantity which, according to the Bayesian, influences one's evaluation of a scientific development is the posterior probability of the revised theory, and for the theory to be 'acceptable' in the everyday sense of the term, this should be relatively high—at any rate, it ought to exceed 0.5. If a theory is more probable than 0.5, then it is more likely to be true than false, which would seem to be a minimum condition for 'acceptability'. On this view, a' will be judged adversely and pejoratively labelled ad hoc, if $P(a' \mid e' \& b) \leq 0.5$, where e' is the new evidence that refuted the predecessor of a' and b is any other relevant information. In this account (which agrees with that given by Horwich, 1982, pp. 105–108), there is no need for a' to be supported by evidence independent of e'; all that is wanted is that it be credible. Scientists are also interested in whether t in the presence of the newly-thought-up a' provides a competent explanation of the previously anomalous e'. It would do so only if $t \& a'$ was a sufficiently credible theory; since $P(t \& a' \mid e' \& b) \leq P(a' \mid e' \& b)$, this would be the case only if a' were not ad hoc.

The Bayesian account explains the low esteem which ad hoc hypotheses frequently command in the scientific community. It also explains why people often respond with immediate incredulity, indeed derision, to an ad hoc hypothesis. Is it likely that their amusement comes from perceiving that the hypothesis leads to no new predictions? We do not believe so. Finally, the Bayesian account explains why the hypotheses are termed ad hoc. For since an ad hoc hypothesis was originally improbable, it would not have been seriously entertained if e', the evidence that undermined an earlier hypothesis, had not been discovered, and the need to explain the new anomalous result had not arisen.

J.4 Why the Standard Account Must Be Wrong

The standard account characterises ad hoc hypotheses as being unsupported or unsupportable by evidence independent of that which led to their being proposed. It is implicit in that account that only hypotheses that do enjoy such support are acceptable. We shall argue that this is wrong, both in the light of counter-examples and by means of a more general argument. The standard account of ad hocness is, no doubt, inspired in part by the desire to avoid attributing inevitably subjective probabilities to theories. If so, the aim backfires, for, as we shall show, the non-Bayesian account has its own subjective aspect, one which, in our view, is very inappropriate.

Consider first a couple of counter-examples to the standard account. Suppose one were examining the hypothesis that an urn contains only red counters. An experiment is conducted in which counters are removed at random and then replaced, and this trial is repeated, say, 10,000 times. Let the result of this trial be that 4950 of the selected counters were red and the rest white. The initial hypothesis, and the various necessary auxiliary assumptions, are together refuted, and a natural revision would be that the urn contains red and white counters in approximately equal numbers. This seems a perfectly legitimate procedure, and the revised hypothesis appears well justified by the evidence, yet there is no independent evidence for it: its support comes solely from the evidence which discredited its predecessor (Howson, 1984).

Theorising about the contents of an urn is only a humble form of enquiry, but there is no reason to think that these conclusions do not also hold in the higher sciences. Indeed, the following is a case where they do hold: the assumption is made that two characteristics of a plant are inherited in accordance with Mendel's principles and that each is controlled by a specific gene, the two genes acting independently and being located on different chromosomes. The results of plant-breeding experiments show that a surprising number of plants carry both characteristics, and the original assumption that the genes act independently is revised in favour of a theory that they are linked on the same chromosome. Again, the revised theory would be strongly confirmed and established as acceptable merely on the evidence that stimulated its formulation and without the necessity of further, independent, evidence. (An example of this sort is worked out by Fisher in his *Statistical Methods for Research Workers*, ch. IX.)

We turn now to a more general objection to the idea that hypotheses are acceptable only if corroborated by independent evidence. Imagine a scientist who performs an experiment and observes e', which because it implies the falsity of the prediction e made by t & a, refutes that combination of theories. Suppose a new theory, t & a', is advanced, which is ad hoc in one or other of the two senses that there is either no fresh evidence for a' or no possibility of such evidence. The theory therefore is unacceptable, according to the view we are considering. But this cannot be so. For consider that only part of the observational evidence, namely $\sim e$, is required for the refutation. Now suppose another scientist first contrived an experiment with only two possible outcomes: either e or $\sim e$. Having obtained the latter, he revises his theory to t & a', performs the orthodox experiment, and observes e'. In cases where e' is not implied or made highly probable by $\sim e$, according to the view we are discussing, this new theory would be perfectly acceptable since it is supported by evidence independent of that which refuted its predecessor. But the two experimenters are in precisely the same position as regards the available evidence, yet for the one the theory is unacceptable, for the other it is not! This is a most alarming consequence for any methodology, for it fails spectacularly to reflect scientific reasoning and flies in the face of common sense.

It is curious, too, that a methodology designed to provide purely objective criteria should lead to the conclusion that the epistemic value of a theory is so closely connected with the state of mind of its inventor.

J.5 The Notion of Independent Evidence

As we have explained, the non-inductivist, non-Bayesian account of ad hocness asserts that a theory consisting of the combination t & a is only replaced by t & a' in an acceptable scientific fashion when a' is successfully tested by evidence independent of that which refuted the first theory. This thesis is often associated with another, rather similar, view, namely, that no theory is acceptable unless it is supported by evidence independent of that which prompted its initial proposal, whether this also refuted a predecessor or not. We have shown that neither of these views is either reasonable or compatible with scientific practice, and, moreover, that they fail to deliver the objective standards of theory-appraisal to which they aspire. (Howson, 1984, addresses a number of other objections.) One problem with the non-Bayesian criterion of ad hocness, which we have not needed to exploit in our criticism of it, is that the notion of 'independent' evidence is left vague and intuitive. Moreover, there seems to be no way of interpreting the notion in a purely objective fashion.

It seems not to be the probabilistic sense of independence that is intended. For suppose $P(e_2 \mid e_1) < P(e_2)$. This means that e_2 is not probabilistically independent of e_1. Nevertheless, if $P(e_2 \mid e_1)$ were sufficiently small, e_2 would (it is generally acknowledged) support an appropriate theory, even if e_1 were already known and had been counted in support of the theory. Hence, the notion of independence that is often employed in this context cannot be the probabilistic notion. Another possibility would be that e_1 is independent of e_2 just in case neither entails the other. But this would mean that if the two bits of evidence were trivially distinct in, say, relating to different times, or slightly different places, then they would be independent. And this would mean that practically no theory would be ad hoc. For instance, Medawar's peculiar theory about the cosiness of the Eskimo's way of life was propounded in response to some surprising IQ measurements that had been reported. Presumably, one could infer from the theory that IQ tests applied the following week to the same group of Eskimos would produce similar results. But although this prediction is logically independent of the earlier reported results, it would not significantly improve the standing of Medawar's theory.

What seems to be wanted of evidence in the standard account for it to save a hypothesis from ad hocness is that it should be supported by evidence that is *different* from that which led to its predecessor's downfall. Indeed, the ideas of dependence and independence seem closely related to those of similarity and diversity, so we shall continue the discussion by considering these notions.

Evidence that is varied is often regarded as offering better support to a hypothesis than an equally extensive volume of homogeneous evidence. As Hempel put it, "the confirmation of a hypothesis depends not only on the quantity of the favorable evidence available, but also on its variety: the greater the variety, the stronger the resulting support" (1966, p. 34). According to the Bayesian, if two sets of data are entailed by a hypothesis (or have similar probabilities relative to it) and one of them confirms that hypothesis more than the other, this must be due to a corresponding difference between the data in their probabilities. In other words, the variety of evidence is a matter of its probability. We shall explain.

Consider first some examples: the report of the rate at which a stone falls to earth from a given height on a Tuesday is similar to that relating to the stone's fall on a Thursday, say; it is very different, however, to a report of the trajectory of a planet or one of the manner in which a given fluid rises in a capillary tube; each of these reports, however, confirms Newton's theory, though to varying degrees. The similar instances in the above list have the characteristic that when one of them is known, any other would thereby be anticipated with high probability. This recalls Francis Bacon's characterisation of similarity in the context of inductive evidence. He spoke of observations "with a promiscuous resemblance one to another, insomuch that if you know one you know all" and was probably the first to point out that it would be superfluous to cite more than a small representative sample of such observations in evidence (*see* Urbach, 1987, pp. 160–164). The idea of similarity between items of evidence is expressed naturally in probabilistic terms by saying that e_1 and e_2 are similar if $P(e_2 \mid e_1)$ is higher than $P(e_2)$ and one might add that the more the first probability exceeds the second, the greater the similarity. This means that e_2 would provide less support if e_1 had already been cited as evidence than if it was cited by itself.

On the other hand, knowing that one of a pair of dissimilar instances has occurred gives little or no guidance as to whether the other will occur. For example, unless Newton's, or some comparable theory, had already been firmly established, a knowledge of the rate of fall of a given object on some specific occasion would not significantly affect one's confidence that the planet Venus, say, would appear in a particular position in the sky on a designated day. Different pieces of evidence may also have a mutually discrediting effect. An example of this might be the observations of the same constant acceleration of heavy bodies dropped at sea level and the unequal rates of fall of objects released on different mountaintops. Both observations would confirm Newton's laws, but in circumstances where those laws are not already well established, the first set of observations might suggest that all objects falling freely (whether on top of a mountain or not) do so with the same acceleration. In other words, with different instances, say e_3 and e_1, $P(e_3 \mid e_1)$ is either close to or less than $P(e_3)$. Of course, e_3 merely being different from e_1 in this sense does not imply that it supports any hypothesis significantly; whether it does or not depends on its probability. The notion of similarity, as we have characterised it, is reflexive, as it should be; that is, if e_2 is (dis)similar to e_1, then e_1 is (dis)similar to e_2 (this follows directly from Bayes's Theorem).

Our characterisation of similarity and diversity in data closely resembles the one we gave earlier in relation to experiments *(see* section **d**). They are not identical, however, since there is nothing in the earlier definitions to preclude the results of one particular experiment counting as a heterogeneous set of data. For an experiment is defined simply in terms of a set of instructions. If such instructions said, for example, 'first throw a die, then if a six appears, measure the speed of light, if a five, ascertain the position of the planet Mars, if a four, . . .'one might obtain varied data, all from the same experiment. It is natural to respond by saying that this complex experiment really comprises a variety of different ones and demand a definition of 'experiment' in terms of the homogeneity of its outcomes rather than in terms of a set of practical directions. Such a definition is no doubt possible, but we feel it is bound to be somewhat arbitrary and unrewarding; it would, for instance, have to stipulate exactly how homogeneous the outcomes of some set of operations should be in order to qualify as a unitary experiment.

■ k INFINITELY MANY THEORIES COMPATIBLE WITH THE DATA

k.1 The Problem

Galileo carried out many experiments on freely falling bodies and on bodies rolling down inclined planes in which he examined how long they took to descend various distances. These experiments led him to formulate the well-known law to the effect that $s = ut + \frac{1}{2}gt^2$, where s is the distance fallen by a freely falling body, u is its initial downward velocity, g is a constant, and t is the time taken by the fall. Jeffreys (1961, p. 3) pointed out that Galileo might also have advanced the following as his law:

$$s = ut + \frac{1}{2}gt^2 + f(t)\,(t - t_1)(t - t_2)\ldots(t - t_n),$$

where $t_1, t_2, \ldots,$ and t_n represent the times at which he carried out his experiments, and where f is any function of t at all. Thus Jeffreys's modification stands for an infinite number of alternatives to Galileo's theory. Although all these theories contradict one another and make different predictions about future experiments, the interesting feature of Jeffreys's unorthodox laws of free fall is that they all imply those data which Galileo had from his experiments.

This is hard to reconcile with those non-inductivist, non-probabilistic theories of scientific method which hold that the scientific value of a theory is determined just by the evidential support it has, where that support is simply a function of $P(e \mid h)$ and, in some versions, of $P(e)$. These philosophical approaches would have to regard the standard law of free fall and those peculiar alternatives described by Jeffreys as equally good scientific theories, relative to the evidence available to Galileo, although this is a judgment with which no scientist would agree. The same point emerges from a well-known example due to Nelson Goodman (1954). He noted that the evidence of very many and varied green emeralds would normally suggest that all emeralds are green. But he pointed out that that evidence bears the same relation to "All emeralds are green" as it does to a type of hypothesis he formulated as "All emeralds are grue". According to Goodman's definition, something is *grue* if it is either green and observed before time t, or blue and observed at or later than t. If t denotes some time after the emeralds described in the evidence were observed, then both the green- and the grue-hypotheses imply that the observed emeralds should be green. However, the hypotheses are incompatible, differing in their predictions about the colour of emeralds looked at after the critical time. As with Jeffreys's variants of Galileo's theory, the grue-hypothesis represents an infinite number of alternatives to the more natural hypothesis, for t can assume any value, provided it is later than now.

Our examples illustrate a general problem for methodology: that a theory which explains (in the sense of implying or associating a certain probability with) some data is merely one out of an infinite set of rival theories, each of which does the same. The existence of this infinite set of possible explanations, it will be remembered, spelled ruin for any attempt at a positive solution to the problem of induction (*see* Chapter 1). The problem with which we are concerned here arises because, in practice, scientists discriminate between possible explanations and typically pick out just one, or at any rate relatively few, as meriting serious attention. An account of scientific method ought to explain how and why they do this.

k.2 The Bayesian Approach to the Problem

This has not proved easy. For the Bayesian, the nature of the problem, at least, is straightforward. Moreover, Bayesian theory does not imply that every hypothesis similarly related to the data is of equal merit. Suppose one were comparing two theories in the light of the same evidence. Their relative posterior probabilities are given by

$$\frac{P(h_1 \mid e)}{P(h_2 \mid e)} = \frac{P(e \mid h_1)P(h_1)}{P(e \mid h_2)P(h_2)}.$$

If both theories imply the evidence, then $P(e \mid h_1) = P(e \mid h_2) = 1$. And if, in addition, $P(h_1 \mid e)$ exceeds $P(h_2 \mid e)$, then it follows that $P(h_1)$ is larger than $P(h_2)$. More generally, if two theories which explain the data equally well nevertheless have different posterior probabilities, then they must have had different priors too. So theories such as the contrived alternatives to Galileo's law and Goodman's grue-variants must, for some reason, have lower prior probabilities. Indeed, this is clearly reflected in most people finding such hypotheses quite unbelievable. The problem then is to discover the criteria and rationales by which theories assume particular prior probabilities.

Sometimes there is a clear reason why a theory is judged improbable. For instance, suppose the theory concerned a succession of events in the development of a society; it might perhaps assert that the elasticity of demand for herring is a constant or that all future British prime ministers' surnames will start with the letter T. These theories, which of course could be true, are however monstrously improbable. And the reason for this is that the events they describe are influenced by numerous independent processes whose separate outcomes are improbable. The probability that all these processes will turn out to favour the hypotheses in question is therefore the product of many small probabilities, and so itself is very small indeed (Urbach, 1987b). The question, of course, remains of how the probabilities of the causal factors are estimated. This could be answered by reference to other probabilities, in which case the question is just pushed one stage back, or else by some different process that does not depend on probabilistic reasoning. For instance, the simplicity of a hypothesis has been thought to have an influence on its initial probability. This and other possible determinants of initial probabilities are discussed in Chapter 11.

It is worth mentioning here that the equation given above, relating the posterior probabilities of two theories with their prior probabilities, explains an important feature of inductive reasoning. The scientist often prefers a theory which explains the data imperfectly, in that $P(e \mid h_1) < 1$, to an alternative, h_2, which predicts them with complete accuracy. Thus, even Galileo's data were not in precise conformity with his theory; nevertheless he did not consider any more complicated function of u and t to be a better theory of free fall than his own, even though it could have embraced the evidence he possessed more perfectly. According to the above equation, this is because the better explanatory power of the rival hypotheses was offset by their inferior prior probabilities (*see* Jeffreys, 1961, p. 4).

■ I CONCLUSION

Charles Darwin (1868, vol. 1, p. 8) said that "In scientific investigations it is permitted to invent any hypothesis, and if it explains various large and independent classes of facts it rises to the rank of a well-grounded theory". This is, perhaps, an exaggeration, for not any hypothesis would do; the hypothesis must not be refuted, or substantially disconfirmed, nor should it be intrinsically too implausible. With these provisos, Bayesianism, we suggest, is just such a well-grounded hypothesis as Darwin referred to. As we showed in Chapter 3, it arises from natural and intuitively reasonable attitudes to risk and uncertainty. It is neither refuted nor undermined by any of the phenomena of scientific reasoning. On the contrary, as we have seen, it explains a wide variety of them. So far we have concentrated chiefly on deterministic theories. We shall see in the next and following chapters that the Bayesian approach is no less successful when dealing with statistical reasoning.

PART III

Classical Inference in Statistics

We showed in Chapter 4 how numerous aspects of scientific reasoning can be illuminated by reference to Bayes's Theorem. We confined the discussion there to deterministic theories. As already explained, however, scientific theories are often not deterministic, but are statistical or probabilistic in character. The evaluation of such hypotheses brings no special problems of principle to a Bayesian analysis, the difference between the cases of deterministic and statistical hypotheses being reflected in the term $P(e \mid h)$ which appears in Bayes's Theorem. In the former case, when h entails e, this term equals 1. When h is statistical, $P(e \mid h)$ takes a value equal to the statistical probability which h confers on e, this being an application of the so-called Principal Principle, which is discussed in Chapter 9. Inductive reasoning about deterministic and probabilistic hypotheses is then explained in a uniform fashion in the Bayesian approach, the former merely constituting a special case of the latter.

No such uniform treatment is afforded, however, by the leading non-Bayesian approaches. As a result, a distinct branch of non-Bayesian, or "classical", statistical methodology has grown up since the 1920s. We shall follow the pattern thus established by dealing separately with statistical and deterministic hypotheses. This plan is justified since classical statistical inference is now a sophisticated and widely applied body of doctrine whose challenge to Bayesianism requires an answer. Moreover, the strength of the Bayesian approach as a unified and successful account of scientific reasoning can best be appreciated when contrasted with what we hope to show are the inadequacies of its main rivals.

In this part of the book we shall review the major aspects of classical statistical inference. These can be divided roughly into two. First, the theory of significance tests, which purports to inform us when we ought to reject a statistical hypothesis or regard it as false. The two rival versions of significance testing were put forward by Fisher, whose contribution is assessed in chapters 5 and 6, and by Neyman in collaboration with Pearson; the Neyman-Pearson account is presented and criticized in Chapter 7. Secondly, estimation theory is an attempt to arrive at a positive conclusion about the so-called best approximation to a true parameter value. This also has two aspects: point estimation and interval estimation. Chapter 8 takes up the subject of the classical theory of estimation.

CHAPTER 5

Fisher's Theory

■ a FALSIFICATIONISM IN STATISTICS

Theories seriously entertained by scientists at one time are often later rejected when reviewed in the light of new experimental evidence. The most straightforward form for such rejections is that of a logical refutation, and provided one is prepared to concede certainty to the refuting data, such refutations may be regarded as scientific modes of inference which require no concession to Bayesian principles. Indeed, some philosophers, keen to avoid a subjective probabilistic assessment of hypotheses, maintain that logical refutations are the only significant type of inference in science. However, as was explained earlier, a large part of modern science is concerned with statistical hypotheses and these are generally not refutable in this way. As an example of a simple statistical hypothesis, take the theory that a particular penny has an even chance of landing heads and tails (the penny is then said to be a 'fair' coin). This theory cannot be refuted by observing the outcomes of trials in which the penny is tossed; no proportion of heads in any sequence, however large, is precluded by the theory. Nevertheless, scientists do not regard statistical theories as necessarily unscientific, nor have they dispensed with procedures for rejecting them in the face of what they take to be unfavourable evidence.

A response to this difficulty that is occasionally canvassed is to attribute a quasi-refutability to statistical theories, whereby, as Cournot (1843, p. 155) expressed it, events which are sufficiently improbable "are rightly regarded as physically impossible". Popper had the same idea: scientists, he said, should make "a methodological decision to regard highly improbable events as ruled out—as prohibited" (1959a, p. 191). And he talked of hypotheses as having been "practically falsified" if they attached sufficiently low probabilities to events that actually occurred. And Watkins endorsed the idea recently as a "non-arbitrary way of reinterpreting probabilistic hypotheses so as to render them falsifiable" (1984, p. 244).

But very improbable events occur. Indeed, it would be difficult to name a probability so small that no event of some smaller probability had not already taken place or is not taking place right now: events of miniscule probability are ubiquitous. Even a probability of $10^{-(10^{12})}$, which Watkins considered to be "vanishingly small" and to amount to an impossibility (1984, p. 244), is nothing of the sort. The probability of the precise distribution of genes in the five billion members of contemporary humanity is incomparably smaller than this, relative to Mendel's laws of inheritance, as is the probability that the atoms in a particular jug of water have a particular spatial distribution at any given time.

Why should anyone think that very improbable events are physically impossible or that they ought to be regarded as such? Popper argued that extremely improbable events that did happen "would not be physical effects, because, on account of their immense improbability, *they are not reproducible at will*" (1959a, p. 203). This unreproducibility of very improbable events, Popper reasoned, means that a physicist "would never be able to decide what really happened in this case, and whether he may not have made an observational mistake". But unreproducible events are not necessarily so fleeting as to prevent a close examination, and Popper's criterion of a physical effect would exclude most natural phenomena from that category since most of them are not under human control; hence, the criterion is misguided.

We shall now describe the rule which Popper presented for fixing an approximate lower bound for permissible probabilities, following Watkins's recent exposition of it. Consider a repeatable experiment, one of whose outcomes occurs m times in n trials. Suppose the ratio $\frac{m}{n}$ lies outside the range $r \pm \delta$ with probability ϵ. Popper's rule enjoins us now to fix δ by reference to the precision with which $\frac{m}{n}$ is measured. We are then required to consider a "large increase" in δ and to ask whether, for a given n, this produces a significant variation in ϵ. If such a change in δ makes "virtually no difference" to ϵ, then ϵ is the probability of an event that should be treated as impossible. This rule seems hopelessly vague, since the distinction between a significant difference in ϵ and "virtually no difference" is undefined, as is the idea of a large change in δ, and both seem inevitably to be matters of taste or arbitrary stipulation. (Watkins, [1984, pp. 244–246] held that an increase in δ by a factor of 40 from 0.1 to 4 per cent of r is a large change, and that the truly *astronomical* factor by which ϵ is diminished from $10^{-(10^{12})}$ to $10^{-(10^{15})}$ represents virtually no change, but he did not say why.) However, even if these vague notions could be standardized, the rule has no epistemic foundation and is accompanied by no reasons why one should apply it.

There is in fact a good reason why one should not. Suppose only heads appeared in a sequence of 1,000 tosses of a coin; then one would normally reject the hypothesis that the coin is fair—a hypothesis which makes such an outcome very improbable indeed. It may be tempting to diagnose this response in terms of Cournot's rule. But if the coin is fair, every outcome of such a trial is equally improbable, for each consists of a particular sequence of heads and/or tails, and so each has a probability of 2^{-1000}. Suppose a critical probability in excess of this minute value had been adopted in compliance with Cournot's rule; this would mean that every outcome should be regarded as physically impossible, which clearly it should not be, since some outcome will certainly occur. If the critical probability were made smaller than 2^{-1000}, then the same consequence follows, though now for some correspondingly larger trial. In other words, whatever probability value is stipulated as dividing physically possible from physically impossible events, there is some perfectly feasible experiment all of whose outcomes would be judged impossible. But this means that if one applied Cournot's rule, statistical hypotheses concerning repeatable occurrences would be treated as if they were certainly false. This is, to say the least, an inadequate view of statistical hypotheses, and it means that the Cournot rule is radically unsound.

■ b FISHER'S THEORY

The idea that falsificationism can be extended to the statistical realm by treating improbable events as if they were impossible is, then, untenable and nowadays does not enjoy much support.

This should not be surprising, since one of the distinctive features of statistical hypotheses is precisely that they do not rule out those events that they class as improbable—they merely specify how unlikely they are. It is, for example, one of the interesting characteristics of the statistical kinetic theory that it opens up the possibility, which had previously been considered closed, that ice can form spontaneously in a hot tub of water. Such an event is given an immensely small probability of actually happening, to be sure, but, nevertheless, unlike other theories, it does not preclude it altogether.

Fisher devised a methodology which was inspired by a quasi-falsificationist view, but which avoids the implication that statistical theories must necessarily be rejected, whatever the outcome. Roughly speaking, Fisher held that a statistical hypothesis should be rejected by any experimental evidence which, on the assumption of that hypothesis, is *relatively* unlikely, relative, that is, to other possible outcomes of the experiment.

The essence of Fisher's idea can be explained with a simple example which we shall frequently find useful. Consider a coin which is tossed in the standard fashion so that each throw is, so far as we can tell, independent of any other. And suppose we are entertaining the fair-coin hypothesis, namely that the coin has a physical probability, constant from throw to throw, of $\frac{1}{2}$ of landing heads and the same probability of landing tails. Fisher called the hypothesis under examination the "null hypothesis". Suppose the coin were tossed 20 times. The first step of a Fisherian analysis would be to specify the range of possible results, the so-called outcome space of the trial. In the present case, this would normally be considered to comprise all the sequences of 20 elements, each of which is either heads or tails— there are 2^{20} such sequences. (In Chapter 7 we shall consider the assumptions underlying any such specification of the outcome space.)

The result of a coin-tossing trial would not normally be stated as a point in the above-mentioned outcome space but would be summarized in some way, usually as the number of heads or tails which it contains. Such a numerical summary, when used in a test of some hypothesis, is known as a test-statistic. We shall discuss the basis upon which test-statistics are chosen in the next section. For the present, we shall work with the number of heads appearing in the outcome as the test-statistic.

The next step is to evaluate the probability of each possible value of the test-statistic on the assumption of the null hypothesis. In general, as we showed in Chapter 2, if the probability of getting heads in a coin-tossing experiment is p and that of getting tails is q, and if separate outcomes are independent, then the probability of n tosses resulting in r heads is given by $^nC_r p^r q^{n-r}$. (See Chapter 2, section **i.**) If $p = q = \frac{1}{2}$, as stated by the null hypothesis in our example, then the probability of r heads occurring in n tosses is $^nC_r \left(\frac{1}{2}\right)^r \left(\frac{1}{2}\right)^{n-r} = {}^nC_r \left(\frac{1}{2}\right)^n$. In the present case, $n = 20$.

We are now in a position to calculate the probabilities associated by the null hypothesis with each possible number of heads that could be obtained in the experiment; the probabilities are listed in the following table and the diagram displays them graphically.

TABLE I The Probabilities of Obtaining r Heads in a Trial Consisting of 20 Throws of a Fair Coin

Number of heads (r)	Probability	Number of heads (r)	Probability
0	9×10^{-7}	11	0.1602
1	1.9×10^{-5}	12	0.1201
2	2×10^{-4}	13	0.0739
3	0.0011	14	0.0370
4	0.0046	15	0.0148
5	0.0148	16	0.0046
6	0.0370	17	0.0011
7	0.0739	18	2×10^{-4}
8	0.1201	19	1.9×10^{-5}
9	0.1602	20	9×10^{-7}
10	0.1762		

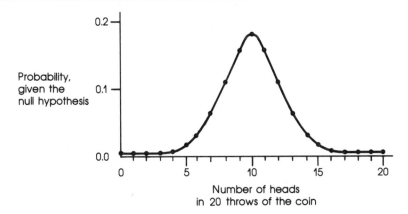

Probability, given the null hypothesis

Number of heads in 20 throws of the coin

According to Fisher's method, the decision to reject the null hypothesis or not is made by noting all the possible results (considered as values of the test-statistic) whose probability is less than or equal to the probability of the result that occurred, these probabilities being calculated on the basis of the null hypothesis. The total probability of obtaining any one of these results, which is the sum of their separate probabilities, must then be calculated. Suppose it takes the value P^*. A convention has grown up, following Fisher, that the null hypothesis should be rejected just in case the value of P^* derived from an experiment does not exceed 0.05. However, some statisticians recommend 0.01, or even 0.001, as the appropriate critical probability. The set of outcomes whose occurrence would lead to the rejection of the hypothesis is called the *critical region*. The value adopted for the critical probability is known as the *significance level* of the test. An outcome falling in the critical region is said to be *significant,* and if such an outcome occurs, then the null hypothesis is said to be *rejected at the 0.05 (or 5 per cent) level*. A test permitting a conclusion of this kind is known as a *significance test*.

Suppose the coin on being flipped 20 times produced 4 heads and 16 tails. Ought the fair-coin hypothesis now to be rejected? In Fisher's method, the question is decided as follows. First, the probability conferred on the result by the null hypothesis is calculated; in the present case (*see* Table I) this is 0.0046. One must now consider those results which could have occurred and which are at least as improbable, given the null hypothesis, as the actual result. Those results are represented in the extremities, or tails, of the above graphical display. The probability of any result falling in these tail-regions is the sum of the probabilities of the outcomes which they cover:

$$P^* = 2 \times (0.0046 + 0.0011 + 2 \times 10^{-4} + 1.9 \times 10^{-5} + 9 \times 10^{-7}) = 0.012.$$

Since $P^* < 0.05$, the hypothesis should be rejected at the 5 per cent level. If, on the other hand, the experiment had resulted in 6 heads and 14 tails, P^* would have been 0.115, which is not significant, and hence the hypothesis should not be rejected at the 5 per cent level.

This simple example illustrates the bare bones of Fisher's approach. It is, however, not always so easy to apply in practice. Take the problem often treated in statistics textbooks of testing whether two populations have the same mean; for instance, whether two particular groups of school children have the same mean IQ. It may be too costly or inconvenient to take measurements from each child, and in such cases a typical experiment would involve selecting a random sample of children from the two groups and evaluating each's IQ. The difficulty in this example is that the null hypothesis does not specify the distributions of IQs in the two populations; it merely asserts that their means are the same. Consequently, no probability distribution can be evaluated for the patterns of IQs which might be found in random samples from those populations, without which no test of significance can be performed. This difficulty is resolved by selecting an aspect of the result which does have a determinable probability conferred on it by the hypothesis. Such aspects are rare and are difficult to find. However, if the samples are normal, which would be the case if sufficiently large samples were drawn from large enough populations, the so-called t-statistic, discovered by W. S. Gossett (who wrote under the name 'Student') has the appropriate qualities for the present example. The Fisherian significance test then takes the usual form. The null hypothesis is stated and the value of t implied by the experimental result is calculated. The known probability distribution of t enables one to establish how probable it is for a result (expressed as a value of the t-statistic) as likely or less so to have arisen. If this probability falls below the previously designated significance level, then the hypothesis is said, as before, to be rejected at the corresponding level.

The essence of Fisher's approach, then, is that rational conclusions about the truth of a hypothesis may be obtained by considering the probability of the result of some repeatable trial, given that hypothesis, relative to the probability of other possible results. In criticizing this approach, we shall first consider what the conclusion of a significance test (that is, the rejection or non-rejection of a hypothesis) amounts to, and whether such conclusions are indeed rational. We shall then discuss the problem posed for the Fisherian method by the fact that the result of a trial can be described in many different ways, not all of them leading to the same conclusion.

■ c DOES FISHER'S THEORY HAVE A RATIONAL FOUNDATION?

The expression 'rejected at such-and-such significance level' is a technical expression which simply records that an experimental result is included in the set of results previously designated the critical region. Describing a hypothesis as rejected in this sense is not equivalent to taking any particular attitude towards its truth or falsity, but is simply stating a purely analytical fact. The controversial inductive step, which is also the essential novelty of Fisher's method, is that of interpreting rejection in something like the everyday sense of the term. In this section we shall consider the permissibility of such a step.

Fisher seems to have seen in significance tests some surrogate for the process of refutation, and he frequently went so far as to say that a theory can be "disproved" in a significance test (e.g., Fisher, 1947, p. 16), and that such tests, "when used accurately, are capable of rejecting or invalidating hypotheses, in so far as these are *contradicted by the data*" (Fisher, 1935; italics added). If Fisher intended to imply that tests of significance can demonstrate the falsity of a statistical theory, and it is difficult to see what else he could have meant, then clearly he was wrong: the experimental results used in a significance test do not logically contradict the null hypothesis. But Fisher was, of course, aware of this, and when he expressed himself more carefully, his justification for significance tests was rather different. The force of a test of significance, Fisher then claimed, "is logically that of the simple disjunction: *Either* an exceptionally rare chance has occurred, *or* the theory of random distribution [i.e., the null hypothesis] is not true" (Fisher, 1956, p. 39). But in thus avoiding an unreasonably strong interpretation, Fisher plumped for one that is unhelpfully weak, for the significant or critical results in a test of significance are by definition improbable, given the null hypothesis. Inevitably, therefore, the occurrence of a significant result is either a 'rare chance' (an improbable event) or the null hypothesis is false, or both. And Fisher's claim amounts to nothing more than this necessary truth. It certainly does not allow one to infer the truth or falsity of any statistical hypothesis from a particular result. (Hacking, 1965, p. 81, has made the same point.)

Expositions of Fisherian significance tests typically vacillate over the nature of the conclusions that such tests entitle one to draw. For example, Cramér said that when a hypothesis has been rejected by such a procedure, "we consider the hypothesis is disproved" (1946, p. 334). He quickly pointed out, though, that "This is, of course, by no means equivalent to a *logical* disproof". Cramér contended, however, that although a rejected theory could in fact be true, when the significance level is sufficiently small, "we *feel* practically justified in disregarding this possibility" (original italics altered). No doubt this is often true (though, as we shall see in Chapter 7, section **c.3,** it is not invariably true); nevertheless, Cramér supplied no rational basis for such a feeling, nor any grounds for thinking that when the feeling arises, it was generated by the type of reasoning employed in tests of significance.

It is often maintained that one is justified in regarding a hypothesis as false when it is rejected in a significance test, on the supposition that if it were tested repeatedly using a significance level of, say, 0.05, then it would 'in the long run' only be wrongly rejected on around 5 per cent of the testing occasions. This claim is often advanced to justify the Neyman-Pearson theory of testing, and we shall discuss it again when we consider that theory in Chapter 7. Suffice it to record here that Fisher did not regard the claim as a feasible defence, for the reason that the indefinite repetition of a significance test is a fiction. He also pointed out that when drawing statistical inferences, the statistician "gives his mind to each particular case in the light of his evidence and his ideas" and is not interested in what would have happened if other experiments had been tried. Fisher might also have offered the more decisive objection that the defence involves a non sequitur. From the probability of a certain type of event being P, it does not follow that even one event of that type will occur in anyone's lifetime, let alone that the event will be observed on any specific percentage of trials. If such a categorical prediction could be drawn from probability statements, they would be falsifiable, an idea that Fisher, in common with practically every other statistician, rightly rejected. Indeed, this rejection was one of the factors that gave rise to significance tests in the first place.

Significance tests do not always require the null hypothesis to be rejected; in particular, when the hypothesis ascribes a sufficiently high probability to the experimental result, it is not rejected. It is perhaps natural to think that it should then be accepted, but what could acceptance mean in this context? We clearly may not regard a hypothesis that survives a significance test as having been definitely established. Can then some lesser degree of approbation be sustained? Cramér, for example, seems to have thought so. Thus he argued that surviving a test of significance "only shows that, from the point of view of the particular test applied, the agreement between theory and observations is satisfactory. Before a statistical hypothesis can be regarded as practically established, it will have to pass repeated tests of different kinds" (Cramér, 1946, pp. 334–335; *see also* p. 420).

But Cramér did not explain why, nor describe in what sense, the so-called agreement between theory and observation is satisfactory, nor did he elucidate the concept of a hypothesis being "practically established", which though clear enough in Bayesian terms, has no obvious classical meaning. Cramér also gave no account of how several significance tests, which are individually unable to improve the epistemic status of a hypothesis, can manage this in combination. So far as we are aware, none of these problems has been satisfactorily dealt with by classical statisticians.

One or two other defences have been offered for significance tests, and these will be aired in Chapter 7, for they have more usually been deployed in connection with the modification of Fisher's principles due to Neyman and Pearson. Neyman and Pearson denied Fisher's claim that theories could be appraised in isolation, and their version of significance tests introduced the idea that a theory should be tested against, or in the context of, rival theories. However, we shall see that these defences provide no greater protection for the principles underlying significance tests than those already examined. Our conclusion will be that one cannot derive scientifically significant conclusions from the type of information which the Fisher and the Neyman-Pearson theories regard as adequate. We shall also see that in order to match significance testing with scientific practice, extra principles need to be imported, and that these involve the personal judgment of the experimenter or statistician. Hence, the much-vaunted objectivity of the classical approach is spurious and, as we shall argue, its claim to improve on Bayesian methods is nullified.

■ d WHICH TEST-STATISTIC?

Fisher's theory as so far propounded is inconsistent. The problem arises because there are usually many ways of summarizing the experimental outcome of a trial into a statistic, and not all of them lead to the same conclusion when put to use as the test-statistics in a significance test. Hence, one test-statistic may instruct you to reject some hypothesis when another tells you not to.

We may illustrate this as follows. First, consider what a statistic is. In general it is characterised as a type of real-valued function defined on the outcome space. That is to say, a statistic associates a unique number with each possible outcome, thereby describing outcomes in a concise numerical way. A statistic is in fact a random variable, as defined in Chapter 2.

Clearly, outcome spaces give rise to many different statistics. Let us consider three. (1) Suppose, for example, the outcome space consists of the 2^{20} permutations of heads and tails which could result from tossing a coin 20 times. Allow these permutations to be arranged in some arbitrary sequence and label any result obtained in an experiment with a number indicating the position it occupies in that sequence; this assignment of numbers would be a statistic. (2) The outcomes in the outcome space might also be divided into separate groups and these groups suitably numbered. Then every outcome could be labelled with the number of the group to which it belongs. This assignment of numbers would be another statistic. And if all the outcomes in each group exhibited the same number of heads, the statistic would be the familiar one that we used as test-statistic earlier. (3) Suppose the last statistic were modified slightly, so that all the outcomes exhibiting 5 heads and 10 heads were combined, and all those with 14 and 15 heads were similarly combined into a single group. The new groups may be arbitrarily renumbered from one to eighteen, thus giving a third statistic defined on the same outcome space. The last statistic is quite artificial, having no natural meaning, but according to the definition, it is a perfectly proper statistic. It is easy to calculate the probability distribution of the modified statistic (which we shall label t') on the hypothesis that a fair coin was used, for it is almost the same as that given in Table I, where the statistic in question was the number of heads in the outcome. The distribution is presented below (*see* Table II). Next to each value of the t'-statistic, we also give the corresponding number of heads in the outcome in order to allow a direct comparison with Table I.

It will be recalled that previously, when the test-statistic was the number of heads, a result of 6 heads and 14 tails was not significant at the 0.05 level. It is easy to see that with the

TABLE II The Probability Distribution of the t'-Statistic

Value of Statistic (t')	Probability	Value of Statistic (t')	Probability
0 (0 heads)	9×10^{-7}	10 (11 heads)	0.1602
1 (1 ")	1.9×10^{-5}	11 (12 ")	0.1201
2 (2 ")	2×10^{-4}	12 (13 ")	0.0739
3 (3 ")	0.0011	13 (14 or 15 heads)	0.0518
4 (4 ")	0.0046	14 (16 heads)	0.0046
5 (6 ")	0.0370	15 (17 ")	0.0011
6 (7 ")	0.0739	16 (18 ")	2×10^{-4}
7 (8 ")	0.1201	17 (19 ")	1.9×10^{-5}
8 (9 ")	0.1602	18 (20 ")	9×10^{-7}
9 (5 or 10 heads)	0.1922		

modified statistic this result now is significant at that level (the P^* value is 0.049). Hence, one must both reject and not reject the null hypothesis, which is impossible.

Clearly there is a need to restrict test-statistics to some group of permissible ones, all of which lead to a similar conclusion. However, any such restriction must be recommended by more than the mere fact that it produces a consistent result; it must produce the right consistent result, if there is one, and for the right reasons, if there are any.

■ e THE CHI-SQUARE TEST

A striking illustration of the difficulties posed by the multiplicity of test-statistics is the chi-square (or χ^2) goodness-of-fit test, which bulks large in the literature, both as a test that is recommended and one that is employed. Gillies (1973, pp. 209–210) pointed out that this test is widely used for testing isolated hypotheses in the manner advocated by Fisher, even by opponents of Fisher's methodology. Gillies's claim that this vindicates the Fisherian view is, however, wide of the mark.

The χ^2 test is used to test the type of hypothesis that ascribes probabilities to several different events against the numbers of each event that actually occur during a trial. The hypothesis might, for example, concern the objective probabilities of a die landing with different numbers showing uppermost when tossed. Suppose the die were thrown n times and that the numbers of occasions it landed with a six, five, etc., showing were $O_6, O_5, \ldots, O_1$. If P_i is the probability that the null hypothesis ascribes to the outcome i, then nP_i is the expected frequency (E_i) of that outcome. The null hypothesis is tested by the following so-called chi-square statistic: $\chi^2 = \sum \dfrac{(O_i - E_i)^2}{E_i}$, where the sum is taken over all the possible outcomes of the trial.

It is a remarkable fact, discovered by Karl Pearson (and proved, for example in Cramér, 1946, pp. 417–419), that this statistic has a probability distribution that depends on the number of so-called *degrees of freedom* of the test, but is practically independent of the unknown probabilities and of n. In general, if the outcome space is divided into J separate cells in the calculation of χ^2, then $v = J - 1$ is the number of degrees of freedom. The probability density distributions of χ^2, for various values of v, are roughly as follows:

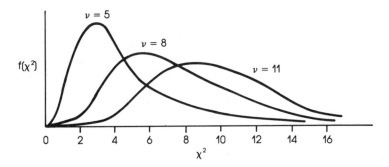

Consider an example. Let the null hypothesis assert that a given die has equal probabilities, of $\frac{1}{6}$, of landing on any one of its sides, and suppose the experiment involves throwing the die 600 times, the following results being obtained:

TABLE III

Outcome i	Observed Frequency O_i	Expected Frequency E_i
Six	90	100
Five	91	100
Four	125	100
Three	85	100
Two	116	100
One	93	100

The expected frequency of each outcome relative to the null hypothesis is equal to nP_i, which in the present case is $600 \times \frac{1}{6}$. Chi-square for the above data is thus given, in accordance with the above formula, by

$$\chi^2 = \frac{1}{100} [(100 - 90)^2 + (100 - 91)^2 + (100 - 125)^2 +$$
$$(100 - 85)^2 + (100 - 116)^2 + (100 - 93)^2]$$
$$= 13.36.$$

Since there are six groups of outcomes, the number of degrees of freedom is five and, as can be roughly gauged from the above sketches, and more precisely established by consulting the appropriate tables, the probability of obtaining a value of chi-square as large or larger than 13.36 is less than 0.05, so the result is significant and the hypothesis must be rejected at the corresponding significance level.

Similar tests are frequently made on theories concerning continuous probability distributions. For instance, the standard method of testing whether a quantity is distributed in a particular fashion, say normally, is to divide the range of possible results of some trial into several intervals and to compare the actual numbers of subjects falling into each with the theoretically expected number, proceeding then as with the simple example of the die.

The chi-square test has been extended as well to cases where the null hypothesis is expressed in terms of undetermined parameters, and where, as a result, χ^2 values cannot be evaluated. The difficulty is often resolved by a method that was proposed by Fisher, in which the unknown parameter values are first estimated from the data and the null hypothesis then restated with the unknown parameters replaced by their estimated values. (The general principles governing classical estimation methods are described in Chapter 8.) Fisher showed that the χ^2 distribution relative to the revised null hypothesis was what it would have been if the parameters had been independently given, but with one degree of freedom fewer for each parameter estimated from the sample. This result, which applies to a wide range of parameter-estimation methods, was proved by Fisher in a mathematical tour de force that occupies nine pages of Cramér's well-known textbook.

Despite the technical sophistication achieved by the chi-square test, it is vitiated, we believe, because no hard-and-fast rule exists for dividing up the outcome space into separate cells, and because whatever method is adopted for effecting a division may influence whether a hypothesis should be accepted or rejected. For instance, reverting to the above example, if only three cells had been formed, say, by combining the pairs of outcomes [six, five], [four, three], and [two, one], the result would then *not* have been significant at the 5 per cent level. The problem may, perhaps, seem even more acute where theories concerning continuous distributions are under test, for such cases present no natural divisions in the range of results, which force themselves on our attention.

This problem is, surprisingly, not usually taken up in expositions of the chi-square test. When it is dealt with, it is always resolved by considerations of convenience rather than of epistemology. For instance, Kendall and Stuart (1979, p. 457) argued that the class boundaries should be drawn so that each cell has an equal probability (as determined by the null hypothesis) of containing the experimental outcome, and they defended this rule on the inadequate grounds that it "is perfectly definite and unique". (The rule was suggested by Mann and Wald, 1942, and Gumbel, 1952, and defended also by Cochran, 1952 and 1954.) However, it is not true that the equal-probability rule produces a unique result. Thus, suppose it had been decided to divide the outcome space into three cells in order to test the null hypothesis that a die is fair. Since there are six possible outcomes, the cells would each contain a pair of these outcomes. The requirement that there be an equal probability of a result falling in any one of the cells does not fix their composition. In fact, there are fifteen ways of constructing cells in line with the equal-probability rule and only two of them produce results significant at the 5 per cent level, when applied to the data in Table III. Secondly, the requirement that each class should have the same chance of including the trial result still leaves open the number of classes to be constructed, a question on which Kendall and Stuart only offer some vague advice, with no assurance that their suggestions have any epistemic basis.

The complacency shown by statisticians in the face of these difficulties is astonishing. Most fail to mention them, and those who do, underestimate their gravity. Thus, although Hays and Winkler warned readers repeatedly and emphatically "that *the arrangement into population class intervals is arbitrary*" (1970, p. 195), their exposition of the chi-square test proceeds as if this fact had no bearing at all on whether the test has any point to it. Kendall and Stuart themselves held that the problem about the formation of classes is "not very serious" (1979, p. 465), but here they were wrong: the difficulty is catastrophic for the chi-square test. Cochran claimed that it is a "minor" problem, merely providing "an argument for more standardization in the application of the test" (1952, p. 335). However, standardization cannot help. It would merely bring about a universal application of pointless and arbitrary principles. What is needed instead is a test based on reasonable principles, and the present problem suggests this is beyond the scope of the chi-square test. It should be discarded.

It might be mentioned, too, that besides the drawback just discussed, the chi-square test also conflicts with various other intuitions about scientific evidence. Suppose, for instance, that the results of the above trial with the die had been as follows:

TABLE IV

Outcome	O_i	E_i
Six	100	100
Five	100	100
Four	100	100
Three	100	100
Two	123	100
One	77	100

If the null hypothesis ascribed the same probability to each of the six faces appearing uppermost in a throw, χ^2 would take the value 10.58, which is not significant at the 5 per cent level. One is, therefore, under no obligation to reject the null hypothesis, even though that hypothesis has pretty clearly got it badly wrong, in particular, in regard to the outcomes two and one. (The same point is made by Good, 1981, p. 161.)

So far we have only considered cases where comparatively large values of χ^2 are taken as grounds for rejecting a hypothesis. But Fisher generally regarded both extremities of a sampling distribution as containing critical values. It is thus consistent with this view to regard small values of χ^2, as well as large ones, as critical. This means, for instance, that if in an experiment the expected and the observed values are the same, or practically so, the null hypothesis could be "as definitely disproved" as if they had been sharply discrepant (Fisher, 1970, section 20). Small χ^2 values are produced by data which we referred to earlier as "too good to be true". Indeed, Fisher's famous criticism of Mendel's experimental results, to which we referred in Chapter 4, section **i,** was based on the fact that they yielded χ^2 values which he claimed were too small to be compatible (in the significance-test sense) with Mendel's genetic theory. Clearly this objection draws on the null hypothesis and on the data alone, for these are the only factors entering into the χ^2 test; whether sufficiently plausible alternatives to the null hypothesis are available plays no role. This means that Fisher's objection would be unaffected even if each of his tentative explanations for the seeming bias in Mendel's results were refuted, a conclusion which, as we argued in the last chapter, is counterintuitive.

Many statisticians have taken a different view to Fisher's. Some would never allow small χ^2 values to discredit a null hypothesis. Others, such as Stuart (1954), maintain that a small χ^2 may be critical, but only if all "irregular" alternatives to the null hypothesis have been excluded. Stuart defined an irregular alternative as one that imputes to the experimental situation "variations due to the observer himself, which include all voluntary and involuntary forms of falsification, however caused, which may make the observations dependent". But he did not say how such a mass-exclusion could be carried out in accordance with classical principles, save to remark cryptically that it would have to "rely on non-mathematical methods". Stuart also did not explain why he only considered the possibility of dependence between the observations arising through some form of interference by the observer, when there clearly could be many natural sources of this effect too.

■ f SUFFICIENT STATISTICS

It is sometimes argued that consistency may be restored to the theory of significance tests by restricting acceptable test-statistics, in particular to those containing all and only the information relevant to the theory under examination. This condition is normally interpreted as meaning that the test-statistic must be a so-called *minimal-sufficient statistic*. We shall, however, argue that this is not a solution available to Fisher. First we must examine the concept of a sufficient statistic.

Some statistics abstract more information from the outcomes than others. For example, a statistic which assigns distinct numbers to each element of the outcome space (such as we described on p. 131) preserves all the information about the outcome, for any value of such a statistic determines the precise element of the outcome space from which it was derived. On the other hand, statistics which map several different outcomes onto a single number thereby discard information. For instance, four coin-tossing trials can result in one of the sixteen sequences of heads and tails (HHHH), . . . , (TTTT). A statistic that records only the number of heads leaves out some information; thus, if the statistic took the value 3, it would not be possible to determine from which of the four different outcomes possessing three heads it was derived. Whether some of the discarded information is relevant to an inference is a question addressed by the theory of sufficiency.

Suppose one were interested in the value of some parameter, say the mean of a normally distributed population or the number of red tickets in an urn. A sample statistic, t, is said to be *sufficient,* relative to such a parameter, just in case the probability that any particular member of the outcome space occurs, conditional on t taking any given value, is the same whatever the parameter's value. Otherwise expressed, if the sample consisting, say, of the n observations, $x_1, \ldots, x_n$ is denoted by $\mathbf{x}$, then t is sufficient for a given parameter just in case $P(\mathbf{x} \mid t)$ is independent of that parameter. And it seems natural to say that $\mathbf{x}$ provides no information about the unknown parameter that is not also in t. Certainly Fisher understood sufficiency this way: "The Criterion of Sufficiency", he wrote, is the rule that "the statistic chosen should summarise the whole of the relevant information supplied by the sample" (Fisher, 1922, p. 316).

Although this interpretation is plausible, it is incompatible with the rest of Fisherian methodology, for according to Fisher's notion of a significance test, two different statistics may both be sufficient and yet lead to contradictory conclusions about whether or not a null hypothesis ought to be rejected. There is a simple example of this in the two statistics we considered in connection with a test on the fair-coin hypothesis: the statistic that preserved all the information about the order in which heads and tails appear in a trial and the statistic defined as the proportion of heads in the trial sequence. Both are sufficient, but while the first never leads to the rejection of a null hypothesis, the second sometimes does. In cases such as these, although the two statistics are sufficient in the technical sense, from the point of view of Fisher's inductive methodology they do not carry the same relevant information; hence, sufficiency is not an adequate criterion of relevancy within the Fisher scheme. According to the Neyman-Pearson and the Bayesian theories it is, though, as we shall show.

Since the sufficiency-condition does not ensure a unique conclusion for Fisherian significance tests, an additional restriction is sometimes suggested, namely that the test should not only be sufficient, but also minimal-sufficient. In other words, the sufficient statistic should be such that any further reduction in its content would destroy its sufficiency. A minimal-sufficient statistic is naturally thought of as containing nothing but relevant aspects of the experimental result.

The minimal-sufficiency condition has been defended recently by Seidenfeld, unsuccessfully, we think. He pointed out that when a deterministic hypothesis is rejected by a refuting instance, there is a certain minimally informative description of the observation which will effect the refutation, and he argued that this is "So too for the statistical case", on the grounds that

> The r.v. [i.e., the random variable used as test-statistic], which is the counterpart for the observation report [in deterministic examples], must make use of only relevant features of the experimental outcome. In other words, the description of the experimental outcome *need* use only relevant properties, relevant to h_o [the statistical hypothesis under test]. (Seidenfeld, 1979, p. 83; original italics replaced)

It is a truism that statistical tests need only use relevant information, but it by no means follows that such tests *must* do so, as Seidenfeld wished to maintain. Indeed, it would be surprising if a case could be made for the minimal-sufficiency proposal, for if information is really irrelevant, it should make no difference to the result of a test, so there should be no need to exclude it.

We have argued that within the Fisherian framework, the sufficiency of a statistic does not imply its inclusion of all the relevant aspects of the data. Hence the attempt to rescue Fisher's significance tests from inconsistency by requiring test-statistics not to leave out any relevant information is bound to fail when this requirement is interpreted in terms of sufficiency. But, in any case, the idea underlying this attempt is almost certainly misconceived, for the reason why different statistics give different results in Fisherian significance tests appears not to be essentially connected with the amount of information they contain. This may be seen by considering an infinite outcome space with a continuous probability density-distribution. As Neyman showed, there may be pairs of statistics based on such an outcome space that are related by a one-to-one transformation, one of which leads to the rejection of a hypothesis (at a specified significance level) while the other does not. Since the two statistics are one-to-one transformations of each other, they necessarily carry the same information, so there must be some other source of the trouble. Neyman (1952, pp. 45–46) traced the problem to the fact that Fisher's tests are carried out on isolated hypotheses; he held that the significance of evidence for a hypothesis turns not only on its probability relative to that hypothesis, but also on its probabilities relative to alternative hypotheses (*see* Chapter 7).

■ g CONCLUSION

Before the 1920s, Bayes's Theorem was widely regarded as the proper basis for inductive reasoning. However, Fisher argued forcefully that it could not deliver completely objective appraisals of hypotheses (we shall answer this charge in Chapter 10), and he led a highly influential and effective campaign against the Bayesian approach. The theory of significance tests was designed as a logically rigorous and fully objective alternative. As we have seen, a part of the theory is perfectly correct from a logical point of view and, hence, uncontroversial. However, the crucial inductive step that prescribes the conclusion either to accept or to reject a hypothesis is questionable. In order to derive a unique conclusion, arbitrary decisions need to be taken, for instance, as to the appropriate test-statistic. These decisions are arbitrary since the various rules recommended for fixing the test-statistic have been given no rational defence, seeming to rest on personal judgments alone. As a leading advocate of Fisherian methods admitted, "There is no answer to [the question 'Which significance test should one use?'] . . . except a subjective one", adding in parenthesis that it was "curious that personal views intrude always" (Kempthorne, 1971, p. 480). Indeed, it *is* curious, when one considers that Fisherian methods arose from a dissatisfaction with the Bayesian approach on account of its supposed subjectivity.

A number of other objections might be made against significance testing, but we have postponed their discussion because they are equally damaging to the Neyman-Pearson theory, a development of Fisher's which we shall deal with in Chapter 7. First, however, we turn to Fisher's influential views on the testing of causal hypotheses.

■ a INTRODUCTION: THE PROBLEM

Thus far we have considered Fisher's methods of testing the kind of hypothesis that attributes a particular value to some parameter. Fisher also applied his principles to the case of causal hypotheses, introducing by the way some important variations with which we shall now deal. The sorts of hypotheses we have in mind concern such questions as whether a certain drug increases the chances of recovery from some illness, whether a new therapy alleviates a particular mental disorder, and whether, and to what extent, a new strain of potato has an inherent capacity to produce a higher crop than an established one. Experiments to test such hypotheses are called clinical or agricultural trials.

Let us pursue the potato example. A simple form of a typical experiment for comparing the yield-capacities of two strains would be the following: a field is divided into pairs of plots, each pair being called a 'block'. One member of each pair is planted with the new strain, while the established variety is grown on the other. The field might look like this:

TABLE V   A Possible Distribution of A- and B-strains of Potato over a Field in an Agricultural Trial

	Plot 1	Plot 2
Block 1	A	B
Block 2	A	B
Block 3	A	B
Block 4	A	B

If one of the strains consistently yielded a greater weight of potato than the other, one might be tempted to infer that this was due to a character intrinsic to the potatoes. However, the more wary will recognise that it might have been caused simply by a difference in the fertility of the two sides of the field, or by some variation in the conditions experienced by the plants in the course of their sowing and growing.

The ideal experiment for deciding between alternative explanations would make the plots of land and the growing conditions absolutely the same, apart from the difference in the type of potato seed. Then, any variation in yields must be due to some innate character of the potatoes. It is occasionally imagined that such ideal experiments are feasible. For example, Feiblman (1972, p. 149) described a controlled experiment as one "in which *all* factors are kept constant except the one under investigation" (our italics). But, of course, it is impossible to keep all but one of the factors constant; as Fisher observed, any two experimental situations necessarily differ in "innumerable" ways (1947, p. 18).

Since the ideal experiment is not feasible, Fisher proposed certain principles of experimental design which he believed would be practicable and which he thought would shield the causal process under scrutiny from those innumerable possible differences which could influence the experimental outcome. The experimental design he suggested has two aspects: control and randomization.

Control. First, treatments are made as alike as possible with regard to those factors which are *known* to influence the outcome. Thus, in comparing the yields of different plant varieties, Fisher advised that they be planted in soil "that appears to be uniform, as judged by the surface and texture of the soil, or by the appearance of a previous crop". Also, the plots on which the varieties are planted should be compact, in view of "the widely verified fact that patches in close proximity are commonly more alike, as judged by the yield of crops, than those which are further apart" (1947, p. 64). Similarly, Fisher (1947, p. 41) recommended that when seeds are sown in plant-pots, the soil be thoroughly mixed before distribution to the pots, that the watering of the pots be equalized, and that precautions be taken to ensure that they each receive the same amount of light. These factors are then said to have been "controlled".

Randomization. The second aspect of classical experimental design is randomization. Its object is to overcome the masking effects of conditions (often called 'nuisance variables') whose influence on the course of the experiment has not been recognised and controlled. Fisher's randomization procedure is generally regarded as a brilliant and effective solution to the problem of nuisance variables in experimental design. For example, according to Kempthorne (1979, pp. 125–26),

> Only when the treatments in the experiment are applied by the experimenter using the full randomization procedure is the chain of inductive inference sound; it is only under these circumstances that the experimenter can attribute whatever effects he observes to the treatment and to the treatment only. Under these circumstances his conclusions are reliable in the statistical sense.

Similarly, Kendall and Stuart (1983, pp. 120–21) held that although Fisher's contributions to statistical theory were remarkable and wide-ranging, "Nevertheless, it is probably no exaggeration to say that his advocacy of *randomization* in experiment design was the most important and the most influential of his many achievements in statistics".

Fisher's enormous influence on experimental design, in this respect, is undeniable. As Kempthorne (1966, p. 17) correctly reported, Fisher's ideas "have been taken over by essentially the whole world of experimental scientists". Experiments which are not properly randomized are frequently written off as seriously deficient or even as entirely useless for the purposes of scientific inference (e.g., by Peto, 1978, pp. 26–27; Gore, 1981; and Altman et al., 1983, p. 1490).

In what follows we shall show that the standard defence of randomization does not work and that the problem of nuisance variables is not solved by randomized designs. In Chapter 10 we shall argue that although randomization may sometimes be harmless and even helpful, it is not a sine qua non; we shall maintain (following Urbach, 1985 and 1987a) that the essential feature of a trial that permits a satisfactory conclusion as to the causal efficacy of a treatment is whether it has been properly controlled.

■ b THE PRINCIPLE OF RANDOMIZATION

Consider first what the principle amounts to. It requires, for example, that in comparing the yields of the two types of potato in an agricultural trial, the plots on which each is grown be selected at random. And when testing the effectiveness of a drug for some condition in a clinical trial, the principle demands that patients should be allocated at random to the test group (who receive the drug) and to the control group (who do not).

This is not the place to discuss the controversial question of what randomness really is; we shall, however, consider this question in Chapter 9, section **b.1.ii.** Suffice it to say here that advocates of randomization regard certain repeatable experiments as sources of randomness. Accordingly, they hold that whether a particular seed is planted on the left or the right side of a field could be decided by the throw of a standard coin or die, the draw of a card from a well-shuffled pack, or whether a radioactive element decays or not in a given time-interval. Random-number tables, which are constructed with the help of such random processes, are often recommended as equally effective for applying the randomization rule. However, it would not be good enough simply to assign experimental units to the various treatments in a way which seems haphazard but which is not objectively random in the sense just specified.

■ c THE CLASSICAL JUSTIFICATION OF RANDOMIZATION

Fisher's main argument for randomizing is that it is needed in order to perform a significance test, which he held to be the proper tool for analysing a trial result. In general, Fisher's idea was that a statistical experiment should test a null hypothesis. But this may not always be statistical in form, in which case the statistical element is introduced by randomizing, and since the probabilities associated with the randomizing device are objective and known, it follows (so Fisher believed) that the probability distribution of the experimental outcomes, conditional on the null hypothesis, may be correctly computed. Only then can a significance test be performed. In Fisher's words,

> The full procedure of randomization [is the method] by which *the validity of the test of significance may be guaranteed* against corruption by the causes of disturbance which have not been eliminated [by being controlled]. (Fisher, 1947, p. 19; italics added)

This is Fisher's oft-repeated justification for randomizing, and it is the one embraced by many classical statisticians.

We may illustrate Fisher's argument through the example we have been considering. Suppose that, in reality, the potato varieties, A and B, have exactly the same genetic growth characteristics and let this be designated the null hypothesis. Imagine too that one plot in each block is more fertile than the other and that, this apart, no relevant difference exists between the conditions which the seeds and the resulting plants experience. If pairs of different-variety seeds are allocated independently and at random, one to each of two plots, then there is a probability of exactly a half that any plant of the first variety will exceed one of the second in yield, even if no intrinsic difference exists. The probability that r out of n pairs show an excess yield for A can then be computed and a standard test of significance applied. Fisher (1947, pp. 41–42) gave almost the same example as this.

Before discussing Fisher's case for randomization, we should note that it applies only to a relatively small number of the trials that are actually conducted, for most trials are interpreted not by means of significance tests, but by some process of parameter estimation. For instance, in typical drug trials experimenters most commonly record the proportions of recoveries, or whatever response is being looked for, in the test and control groups, and they then regard these as estimates of a corresponding value that applies to a wider population. (A quick survey of any medical journal confirms this.) We shall consider classical methods of estimation in detail in Chapter 8; it may be noted here that, as with significance tests, classical estimation bases its conclusions on a sampling distribution, that is, on the probability distribution of all the possible outcomes of an experiment. But the sampling distributions employed to reach such estimates in clinical trials are fixed by randomly selecting subjects from the reference population, *not* by allocating them at random to the test and control groups. So Fisher's argument in support of randomization has a rather limited application. Worse still, as we shall see, the argument is anyway invalid.

■ d WHY THE CLASSICAL JUSTIFICATION DOESN'T WORK

Fisher's justification for randomizing assumes (continuing with our example) that by selecting at random the plot of land on which the seeds are sown, one is *guaranteeing* an objective probability of a half that a difference in yields in the predicted direction will occur, even if the two kinds of seed have identical innate growth characteristics. However, the justification proceeds from what seem to be unjustifiably restrictive assumptions. For example, one has to assume that none of the innumerable environmental variations that emerge after the randomization step introduces a corresponding variation in plant growth. For if one of the plant varieties were selectively subjected to some hidden growth-promoting factor, the probability of there arising a difference in yield might not be a half but could have another, unknown value. This difficulty was, of course, perceived by Fisher from the outset. He proposed dealing with it by randomizing at the very last stage in the experimental procedure:

> the random choice of the objects to be treated in different ways would be a complete guarantee of the validity of the test of significance, if these treatments were the last in time of the stages in the physical history of the objects which might affect their experimental reaction. (1947, p. 20)

This could only express an ideal, however, for it is hard to imagine how one could identify the last stage in the history of an experiment which *might* affect its outcome, let alone how or what one would randomize at that stage. But this difficulty "causes no practical inconvenience", according to Fisher,

> for subsequent causes of differentiation, [subsequent, that is, to the normal randomization step] if under the experimenter's control ... can either be predetermined before the treatments have been randomized, or, if this has not been done, can be randomized on their own account; and other causes of differentiation will be either *(a)* consequences of differences already randomized, or *(b)* natural consequences of the difference in treatment to be tested, of which on the null hypothesis there will be none, by definition, or *(c)* effects supervening by chance independently from the treatments applied. (Fisher, 1947, pp. 20–21)

Fisher seems to be saying here that influences which might have affected the experimental result and which have not been controlled or dealt with through randomizing do not matter, for they are either produced by differences already randomized and so are automatically distributed at random, or they are chance effects that are independent of the treatment and so are, as it were, subject to a natural randomization.

But this overlooks the possibility of influences which are neither of these. Take, for example, the potato trial already discussed and suppose that when planted alone, each of the varieties benefits to a similar extent from the attentions of a certain insect, but that if both were growing in the same vicinity, those insects would be preferentially attracted to one of them. Or imagine that the two kinds of potato plant compete unequally for soil nutrients. Effects such as these (of which infinitely many are possible) would favour one of the treatments at the expense of the other, notwithstanding the earlier randomization.

Fisher's defence also fails to take account of disturbing factors that could operate before the plant seeds were sown. For instance, the different seeds could have been handled by different market gardeners, one of whom had some unwitting effect on subsequent growth; or the sacks in which the seeds were stored may have influenced their future development; etc., etc. No later randomization could compensate or undo any unfair advantage that might be imparted by factors such as these. In order to deal with them in the Fisherian way, a separate randomization would need to be devised for each. But because there are infinitely many such possible sources of error, there must be a matching number of randomizations to guarantee the significance test. As this is impossible, randomization cannot provide such a guarantee. Hence the frequently voiced claim for randomization that it "protects against sources of bias that are *unsuspected*" (Snedecor and Cochran, 1967, p. 110) and Fisher's opinion that it "relieves the experimenter from the anxiety of considering and estimating the magnitude of the *innumerable causes* by which his data *may* be disturbed" (1947, p. 43; our italics) have to be dismissed.

■ e A PLAUSIBLE DEFENCE OF RANDOMIZATION

It might be plausibly argued against these criticisms that many variations in the experimental process, while they are possible sources of error are, however, not likely ones. For example, the colour of the market gardener's socks and the size of his shirt collar are conceivable but scarcely credible as influences on plant growth. And it seems reasonable to say that such factors can be safely ignored in the experimental design. Indeed, this is what is normally said, or at any rate tacitly held, by advocates of randomization. For instance, Kendall and Stuart acknowledged that in designing a trial, one has to evaluate the relative importance of possible extraneous influences on its outcome:

> A substantial part of the skill of the experimenter lies in his choice of factors to be randomized out of the experiment. If he is careful, he will randomize out all the factors which are suspected to be causally important but which are not actually part of the experimental structure. But every experimenter necessarily neglects some conceivably causal factors; if this were not so, the randomization procedure required would be impossibly complicated (1983, p. 137).

In accordance with these sentiments, when examining the effects of various doses of alcohol on a person's reaction time, Kendall and Stuart explicitly omitted the colour of the subjects' eyes from any randomization, since this is "almost certainly negligible" as an influence on the effect being studied. No experimental data were cited to support this claim, presumably because none exists. But even if a careful trial had been made to investigate the effects of eye colour on reaction times, certain conceivable influences on *its* outcome would have to have been set aside as negligible. Any demand that an influence be ignored as "almost certainly negligible" only after a properly randomized trial would, therefore, be futile, since it would lead to an infinite regress. Presumably for this reason, Kendall and Stuart (1983, p. 137) concluded that the decision whether a factor should be "randomized out" (their phrase) or neglected *"is essentially a matter of judgement"* (our italics).

So according to Kendall and Stuart, one cannot ensure that every possibly relevant factor is randomized in a trial. What one should do, according to them, is discriminate, through the exercise of a personal judgment, between those factors that are worth randomizing and those that are not. This seems to us an inevitable consequence of Fisher's position. However, the example of eye colour is not well chosen by Kendall and Stuart to illustrate the point, for in their experiment, different doses of alcohol are allocated randomly to the subjects, and this means that eye colour and any other characteristic to which the subject is permanently attached is, contrary to their claim, automatically randomized. But this does not detract from Kendall and Stuart's point, for there are many other examples they could have used. For instance, the various doses of alcohol may have been fed to the subjects from different vessels, they will contain unequal volumes of the liquid used to dilute them and different quantities of whatever trace impurities are contained in the alcohol, and so on. None of these factors would be distributed randomly over the different alcohol doses; but this would not matter since, like eye colour, they would presumably all be judged "almost certainly negligible" as disturbing influences.

Kendall and Stuart's claim is that randomization should be restricted to factors that are, in the scientist's judgment, likely to influence the course of the trial.

■ f WHY THE PLAUSIBLE DEFENCE DOESN'T WORK

This modified principle of randomization neither assists nor reinstates Fisher's case. First, it exposes yet another essentially personal element in what purports to be a purely objective account of scientific inference. Secondly, there now seems no point in randomizing. True, it ensures that presumed or suspected variables that have not been controlled will be randomly distributed, but it does nothing for the unsuspected variables, the nuisance of which it was Fisher's aim to remove. Moreover, the question arises why one should randomize at all, for if it was a good idea to control factors 'known' to affect the experimental outcome, then it would appear just as sensible to do the same with factors whose significance is less certain. Indeed, this is just what is almost universally done, though in a roundabout way. For medical researchers are regularly advised always to "check that the groups as randomized do not differ with respect to characteristics as assessed before treatment begins" (Gore, 1981, p. 1959). In other words, before continuing the trial, one should inspect the randomized groups to see whether they display any significant differences; if they do, they should be disbanded and the random allocation started afresh. Even Fisher, in a conversation with L. J. Savage, admitted he would do this: "'What would you do,' I had asked, 'if, drawing a Latin square at random for an experiment, you happened to draw a Knut Vik square?'" (These squares are kinds of chequerboard patterns, in which fields are laid out in agricultural trials.) "Sir Ronald said he thought he would draw again and that, ideally, a theory explicity excluding regular squares should be developed" (Savage, 1962a, p. 88).

Fisher, of course, is not entitled simply to discard products of a random allocation, for this would alter the outcome space of the trial and hence change the sampling distribution. He would have to have specified in advance which configurations were acceptable and which not and recalculated the sampling distribution accordingly. This, so far as we can tell, is never done. But the more significant conclusion is that if all the factors suspected of playing some causal role in the trial could be controlled, and if, as Kendall and Stuart maintained—correctly, in our view—all other factors may be ignored, then randomization is dispensable.

The fact that in practically all randomized trials the experimental groups generated by chance are then carefully inspected to ensure that they are similar in relevant respects suggests strongly that the randomization was not after all essential, and that the real aim in such trials is not to distribute nuisance variables at random, but to compare properly matched groups. This would explain the need to control known and presumed disturbing variables. It would also explain the frequently heard but faulty claim in support of randomization that it ensures or at least makes probable the even distribution over the comparison groups of all other factors whose effects are unsuspected. Gore, for instance, advanced this argument: "randomization", she claimed, "is an insurance in the long run against substantial accidental bias between treatment groups with respect to some patient variable", though she conceded that this "guarantee *(sic)* applies only to large trials", by which she meant ones with more than 200 subjects. This standard claim is, however, quite untrue. Whatever the size of the sample, two treatment groups are *absolutely certain* to differ in some respect, indeed, in infinitely many respects, any of which might, unknown to us, be causally implicated in the trial outcome. So randomization cannot possibly guarantee that the groups will be free from bias by unknown nuisance factors. And since one obviously doesn't know what those unknown factors are, one is in no position to calculate the probability of such a bias developing either.

To summarize. Fisher envisaged an experimental design

which would permit reliable inductive inferences to be drawn, whatever unknown extraneous influences happened to be operating. But as we have seen, unless these possible influences could be assessed as to whether they are likely to have a significant effect, Fisher's recommendations would be ineffective, for they would require infinitely many randomizations. On the other hand, were such an assessment allowed, then the randomization step would not be essential. This is not to say that randomization may not sometimes be a useful way of selecting groups. We are simply claiming that as a universal remedy for the problem of nuisance variables, randomization is ineffectual. We conclude, therefore, that randomization should not be regarded as a sine qua non for agricultural or clinical trials. (The discussion is taken up again in Chapter 10, section **f.**)

■ g FURTHER OBJECTIONS TO RANDOMIZATION

The randomization requirement may seem too trifling to be worth contesting. In fact, however, it imposes severe and, in our view, undesirable limitations on experiments.

Thus the principle of randomization excludes as illegitimate so-called historical controls. Historical controls suggest themselves when, for example, one wishes to find out whether a new therapy increases the chance of recovery from a particular disease, compared with an established treatment. In such cases the new drug would be administered to the test group and the established drug to the control group. Since many patients would already have been observed under the old regime, it seems unnecessary and extravagant to submit a new batch to that same treatment. The control group could be formed from past records and the new treatment applied to a fresh set of patients who have been carefully matched with those in the artificially constructed control group (or historical control). But the theory of randomization prohibits this kind of experiment, since patients are not assigned with equal probabilities to the two groups; indeed, subjects finding themselves in either one of the groups would have had no chance at all of being chosen for the other.

There is also sometimes an unattractive ethical aspect to randomization, particularly in medical research. A new treatment which is deemed worth the trouble and expense of an investigation has often recommended itself in pilot studies and in informal observations as having a reasonable chance of being better for the patient than established methods. But if there were evidence, however modest, that a patient would suffer less with the new therapy than with the old, it would surely be unethical to expose randomly selected sufferers to the established and apparently inferior treatment. Yet this is just what the theory of randomization insists upon. No such ethical problem arises when patients receiving the new treatment are compared with a matched set who have already been treated under the old regime.

The principle of randomization also represents an implausible departure from normal practice in science. Physicists, for example, do not conduct experiments as Fisher would have them do and yet they are often interested in the same type of question, namely whether some treatment has a causal effect or not. Take a simple experiment to measure the acceleration due to gravity in which a heavy object is dropped close to the Earth. The conditions would be controlled by ensuring that the air is still, that the space between the object and the ground is unimpeded, and so on for other factors likely to interfere with the rate at which the object descends. What no scientist would do is divide the Earth's surface into small plots and then select some of these using random-number tables for the places to perform the experiments. Randomizers might take one of two attitudes to the non-randomizing preference of physical scientists. They could either say it is irrational and ought to be given up or else insist that experiments in physics and chemistry are, in some crucial respect, unlike those in biology and psychology, neither of which seem promising lines of defence.

■ h CONCLUSION

We have argued that randomization does not solve the problem for which it was designed, and, moreover, that there are good reasons for not regarding it as an absolute precondition on trials. In Chapter 10 we shall argue that a reasonable treatment of the problem of nuisance variables and intuitively correct principles of experimental design are provided through the Bayesian approach.

CHAPTER 7

The Neyman-Pearson Theory of Significance Tests

■ a AN OUTLINE OF THE THEORY

The two problems that we have examined and have seen to undermine Fisher's theory are, first, that it offers no satisfactory rationale for the procedures it recommends, and secondly, that in any case, those procedures are incomplete, insofar as they leave open which test-statistic should be used. Neyman and Pearson (1928) proposed a theory which seemed to resolve both these difficulties. The novelty in their approach was to test a statistical theory not in isolation, but in the context of, or against, rival theories. This marks a significant departure from Fisher's idea that statistical theories should follow deterministic ones in being capable of rejection without regard to the performance of any other theory. However, with Fisher, Neyman and Pearson accorded no role to prior or posterior probabilities of theories, being convinced that they could create a perfectly objective, non-Bayesian method of statistical inference.

In expounding the Neyman-Pearson method, we shall first consider cases where there are just two competing hypotheses, h_1 and h_2. Neyman-Pearson tests, like those of Fisher, allow for only two kinds of inference: either a hypothesis is rejected or it is accepted. Such inferences are subject to two sorts of error: one could regard h_1 as false when in fact it is true or accept h_1 (and, hence, reject h_2) when it is false. When these errors can be distinguished by their gravity, the more serious of them is normally called a *type I error* and the less serious a *type II error*. The seriousness of the two types of error is judged by the practical consequences of acting on the assumption that the rejected hypothesis is false and the accepted one true. For example, suppose two alternative theories concerning a proposed food additive were entertained, one that the substance is safe, the other that it is highly toxic. Under a wide variety of circumstances, it would be less of a danger to assume that a safe additive was toxic than that a toxic one was safe. When the erroneous rejection of a hypothesis is classified as type I, Neyman and Pearson, adopting Fisher's term, called it *the null hypothesis*. If the errors seem equally serious, then it is a matter that may be decided arbitrarily which occupies the role of null hypothesis. The aim of the Neyman-Pearson approach is to minimize the chance of committing both types of error. The possibilities for error are summed up in Table VI.

TABLE VI

Decision	True Hypothesis	
	h_1	h_2
Reject h_1	Error	/
Accept h_1	/	Error

Let us consider the Neyman-Pearson approach through an example, which we have borrowed from Kyburg (1974, pp. 26–35). A purchaser of tulip bulbs cannot remember whether he ordered a consignment containing 40 per cent red-flowering and 60 per cent yellow-flowering tulips or 60 per cent red- and 40 per cent yellow-flowering plants. We shall label these possibilities h_1 and h_2, respectively, and treat the former as the null hypothesis. An experiment to test these hypotheses might involve planting, say, 10 bulbs that have been randomly selected from the consignment and observing which grow red and which yellow flowers. The testing procedure is similar to Fisher's, as described in Chapter 5, and involves the following steps.

First, the outcome space of the experimental sampling must be specified and in the present case, it may be considered to be the set of sequences, each containing as many elements as the sample, each element representing either a red- or a yellow-flowering plant. Thus each sequence states the flower-colour of the tulip bulb selected first, then of the one selected second, and so on down to the tenth. Clearly, there are 2^{10} such sequences. Secondly, a test-statistic needs to be stipulated, in terms of which the outcomes may be summarised in numerical form. We shall work with the number of red plants appearing in the sample as the test-statistic. (These somewhat arbitrary-seeming decisions about the outcome space and the test-statistic will be reviewed later in the chapter.) Next, it is necessary to compute the probabilities of each possible outcome relative to each of the rival hypotheses. If, as we shall assume, the consignment of tulip bulbs is large, the probability of selecting r red-flowering bulbs in a sample of n is approximated by the familiar binomial term ${}^nC_r p^r (1 - p)^{n-r}$, where p is the probability of selecting a red-flowering bulb. (We are, then, assuming that this probability remains constant from selection to selection, an assumption which is approximated the more closely the larger the population of bulbs.) Using the binomial formula, the probabilities of each possible outcome of the experiment can now be calculated, first of all, on the assumption of h_1, that $p = 0.4$, and, secondly, on the assumption of h_2, that $p = 0.6$. The results of these calculations, based on a sample size of 10, are presented in Table VII:

TABLE VII

Outcome (Red, Yellow)	h_1 (p = 0.4)	h_2 (p = 0.6)
0, 10	0.0060	0.0001
1, 9	0.0403	0.0016
2, 8	0.1209	0.0106
3, 7	0.2150	0.0425
4, 6	0.2508	0.1115
5, 5	0.2006	0.2006
6, 4	0.1115	0.2508
7, 3	0.0425	0.2150
8, 2	0.0106	0.1209
9, 1	0.0016	0.0403
10, 0	0.0001	0.0060

These results can be displayed graphically as follows, with crosses corresponding to h_1 and circles to h_2:

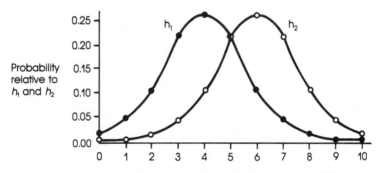

The number of red tulips in a sample of 10

Finally we need a rule that will tell us when to reject the null hypothesis. Suppose the rule were adopted to reject h_1 if 6 or more red-flowering bulbs appear in a sample of 10. If h_1 is true, the probability of the sample containing at least 6 red-flowering bulbs is (consult Table VII, 2nd column)

$$0.1115 + 0.0425 + 0.0106 + 0.0016 + 0.0001 = 0.1663.$$

Hence, according to the suggested rule, h_1 would be rejected with probability 0.1663, if it were true. If h_1 were treated as the null hypothesis, this means that the probability of a type I error would be 0.1663. This probability is called the significance level, as before, or the *size* of the test. The probability of a type II error is the probability of accepting h_1 when it is false. On the assumption that one of the two hypotheses is true, this is identical to the probability of rejecting h_2 when it is true, which is (see Table VII, 3rd column)

$$0.2006 + 0.1115 + 0.0425 + 0.0106 + 0.0016 +$$
$$0.0001 = 0.3664.$$

Hence, the probability of a type II error is 0.3664. The *power* of a test is defined as $1 - P(\text{type II error})$ and is regarded as a measure of the degree to which the test discriminates between the two hypotheses. It is also the probability of rejecting the null hypothesis when it is false. In the present example, the power is 0.6336.

In selecting the size and power of any test, Neyman and Pearson argued that, ideally, one should minimize the former and maximize the latter in order to render the chances of both kinds of error as small as possible. We shall, in due course, consider how well suited this ideal is to the aims of science, but we must first address the more immediate difficulty that the ideal cannot be achieved straightforwardly, for its two aims are incompatible; in most cases, a diminution in size produces a contraction in power, and vice versa. The ideal needs to be modified, and Neyman and Pearson did this with the proposal that one should fix the size of the test at an appropriate level and, within this constraint, maximize its power. Practitioners of the Neyman-Pearson method normally follow the convention instituted by Fisher of setting the test size either at 0.05 or 0.01.

The test employed in the tulip example had a size of 0.1663, which greatly exceeds either of the normally favoured values.

Consider, then, a second test for deciding between the two hypotheses: reject h_1 when 7 or more red tulips appear in a random sample. This test has a size of 0.0548 and a power equal to 0.3823. Note that the power of the earlier test was 0.6336; so while the revised test has a smaller size, this advantage (as most classical statisticians would judge it) is offset by its smaller power.

Randomized tests. It is generally held that the size of a significance test ought not to exceed 0.05 and for a reason we shall report later on, it is often considered advisable for practitioners always to employ roughly the same significance level. But the methods introduced so far cannot be used to construct tests with arbitrarily preselected sizes. For instance, no test of the tulip hypothesis constructed by those methods has a size of exactly 0.10. Tests of any size can, however, be devised by means of so-called randomized tests whose essential features have been lucidly explained by Kyburg in the context of the example we have been discussing. Let the two tests considered above be labelled 1 and 2. As we have shown, they have the following characteristics:

TABLE VIII

	Probability of a type I error	Probability of a type II error	Power
Test 1	0.1663	0.3664	0.6336
Test 2	0.0548	0.6177	0.3823

Consider, now, a third test which is carried out thus: take a well-shuffled pack of 200 cards containing 119 red and 81 black cards and select one at random. If a black card is selected, apply test 1, and if a red, use test 2. This composite test is known as a *mixed* or *randomized* test; it has the desired size of 0.10, this being given by

$$\frac{81}{200} \times 0.1663 + \frac{119}{200} \times 0.0548 = 0.100.$$

The corresponding probability of a type II error is easily worked out to be 0.5159; hence, the power of this test is 0.4841. Randomized tests are rarely, if ever, used, but they form a proper part of the Neyman-Pearson theory. Hence, any criticism which they might merit could quite correctly also be directed at the Neyman-Pearson method in general.

■ b HOW THE NEYMAN-PEARSON METHOD IMPROVES ON FISHER'S

Before examining more complicated and perhaps more realistic examples of testing situations (we do this in section **d** of this chapter), it is worth considering the merits of the Neyman-Pearson approach in the straightforward case we have described, where a theory is tested against a simple alternative. Such an appraisal is justified because, as we shall argue, none of the objections to Neyman-Pearson tests as they apply in elementary situations are in the least mitigated by a more complete treatment. Quite the contrary.

b.1 The Choice of Critical Region

An advantage of the present theory is that it explains an aspect of Fisher's account which seems to have been incorporated rather arbitrarily, merely in deference to apparent scientific practice; that is, that the critical regions of Fisher's significance tests are concentrated in one or both of the tails of the probability distributions of outcomes. It seemed to Neyman and Pearson that if Fisher were merely interested in seeing whether the outcome of a trial fell in a region of low probability, he could, for example, have chosen for that region a narrow band in the centre of a bell-shaped distribution just as well as a wider band in its tails.

The Fundamental Lemma. In the Neyman-Pearson approach, once the space of outcomes and its associated probability distribution are known, the critical region for a particular null hypothesis h_1 and a rival h_2 is uniquely determined for each significance level. This is expressed in the theorem, the proof of which is given in standard texts, that the critical region of maximum power in a Neyman-Pearson test is the set of points in the outcome space that satisfy the inequality:

$$\frac{P(x \mid h_1)}{P(x \mid h_2)} \le k,$$

where k is a constant whose value depends on the hypotheses and on the significance level, and h_1 is the null hypothesis. The probabilities may also be densities. This is known as the Fundamental Lemma of Neyman and Pearson. (Strictly speaking, the likelihoods, $P(x \mid h_1)$, and so forth, should not be expressed here as conditional probabilities, which presuppose what classical statisticians strenuously deny, namely, that the parameter value is a random variable. Hence, they are sometimes written $P(x ; h_1)$ or $L(x ; h_1)$. Bayesians, of course, need have no such qualms.) The Fundamental Lemma embraces cases where a given significance level is achieved by means of a randomized test. In such cases the critical region spoken of in the Lemma comprises the critical regions of the component non-randomized tests. The points in the composite critical region are then selected at random, in accordance with the principles of randomized testing that we have described.

The Neyman-Pearson Lemma fixes uniquely the critical region of a test and thus gives a rationale for some particular region and no other being considered critical. Such a rationale seems to be lacking in the Fisher approach.

b.2 The Choice of Test-Statistic and the Use of Sufficient Statistics

Apart from seeming to set the critical region of a test arbitrarily in the tails (or, sometimes, in just one tail) of the probability distribution, Fisher's theory has the awkward consequence that an experimental result which under one description falls among the relatively improbable outcomes, may not do so when covered by another description.

It will be recalled that in response to this difficulty the idea was put forward that only sufficient statistics should be employed, on the grounds that they alone contain all the relevant information. However, as we showed, the latter claim could not be sustained if understood within the framework of Fisherian tests of significance. According to the Neyman-Pearson account, however, the claim is true; that is, sufficient statistics do hold all the relevant information. This follows from the Neyman-Pearson Lemma. Assume that h_1 and h_2 ascribe different values to a parameter, θ. It will be recalled that where $\mathbf{x} = x_1, \ldots, x_n$ is a point in the outcome space, t is sufficient for θ just in case $P(\mathbf{x} \mid t)$ is the same for all θ. It can be easily shown that if t is sufficient for θ, then

$$\frac{P(\mathbf{x} \mid h_1)}{P(\mathbf{x} \mid h_2)} = \frac{P(t \mid h_1)}{P(t \mid h_2)}.$$

The Neyman-Pearson Lemma asserts that the critical region of maximum power is the set of points in the outcome space for which the left-hand ratio does not exceed some number k. By the above result, the same must hold for the right-hand ratio. Hence, if the outcome space had been specified in terms of the sufficient statistic t, instead of $\mathbf{x}$, the region of maximum power would have included exactly the same outcomes.

■ c SOME PROBLEMS FOR THE NEYMAN-PEARSON THEORY

c.1 What Does It Mean to Accept and Reject a Hypothesis?

It might be said, too, in favour of the Neyman-Pearson method that it is founded on reasonable principles. After all, it seems reasonable to reduce to a minimum the chances of accepting a false theory and of rejecting a true one. However, a closer examination reveals profound objections to this aim.

What, for example, can be meant by 'accepting' and 'rejecting' a hypothesis? Oddly enough, most textbooks in statistics are silent on this crucial question. Yet these terms are far from clear. We can be sure, for instance, that to accept a hypothesis does not amount to regarding it as definitely true, nor can its rejection be equivalent to asserting that it is definitely false, for nobody denies that hypotheses rejected in statistical tests may be true, and that accepted ones may be false.

Moreover, we should not say that hypotheses rejected at a given level of significance, α, thereby acquire a probability of α (or of any other value) of being true, for the Neyman-Pearson theory has this in common with Fisher's, that it accords no role to inductive probabilities. Indeed, as we have pointed out, these theories were developed out of a denial that such probabilities have any scientific meaning.

Another possibility is that on accepting a hypothesis, a scientist behaves *as if* it were true, an attitude manifesting itself, perhaps, in practical actions which presuppose that hypothesis. For example, imagine that a scientist has accepted the hypothesis, discussed above, that a consignment of tulip bulbs contains 40 per cent of the red-flowering variety. According to this interpretation, he would never bother to repeat the experiment. Moreover, he would be happy to stake his entire stock of worldly goods and any other goods he might possess on a wager offered at odds of, say, 10 to 1 against that hypothesis being true. Or suppose a food additive conjectured to be toxic were subjected to a trial involving 10 persons and the conjecture were rejected, then the manufacturer would be prepared to go directly into large-scale production and distribution. This interpretation of acceptance and rejection has merely to be stated to reveal its absurdity. No scientist, nor indeed any rational person, would act in such a crazy fashion. Nevertheless, despite its immense implausibility, this seems to be the way statisticians standardly interpret the notions. For instance, in his text expounding the theory of significance tests, Lindgren (1976, p. 306) assumed that "a problem is one of hypothesis testing when one is willing to take one of two actions—that is, he will *either act as though* . . . [the null hypothesis] *were true, or act as though* . . . [the alternative] *were true*, without necessarily being convinced one way or the other". Similarly, Hodges and Lehmann (1970, p. 311) described the acceptance of the null hypothesis as "The decision to act as if [it] . . . were true".

The idea that one should treat a rejected hypothesis as if it were definitely false and an accepted one as if definitely true is not only standard now but was repeatedly avowed by the founders of the approach we are considering. Thus Neyman and Pearson (1933, p. 142) said that "Without hoping to know whether each separate hypothesis is true or false, we may search for rules to govern our behaviour with regard to them, in following which we will insure that, in the long run of experience, we shall not be too often wrong". It is evident that the behaviour Neyman and Pearson had in mind was the acceptance and rejection of hypotheses as being true or false, that is, the adoption of the same attitude towards them as one would take if one had an unqualified belief in their truth or falsehood. For they defended their position by saying that while their rule "tells us nothing as to whether in a particular case [a hypothesis] h is true", nevertheless,

> It may often be proved that if we behave according to such a rule, then in the long run we shall reject h when it is true not more, say, than once in a hundred times, and in addition we may have evidence that we shall reject h sufficiently often when it is false. (Neyman and Pearson, 1933, p. 142)

It is clear from this that the errors which their approach was designed to mitigate are those of treating a hypothesis as definitely true when it is false, and as definitely false when it is true.

The proof to which Neyman and Pearson alluded and which supposedly justifies the testing procedure they advocated is one we have already criticized as being both irrelevant and invalid. Its irrelevance was highlighted by Fisher, who argued that scientists are not concerned with how rarely they might falsely reject a hypothesis in a series of experiments. As he put it, "To a practical man . . . who rejects a hypothesis, it is, of course, a matter of indifference with what probability he might be led to accept the hypothesis falsely, for in his case he is not accepting it" (Fisher, 1956, p. 42). For this reason, Fisher rightly dismissed the Neyman-Pearson defence as "absurdly academic". Fisher failed, however, to point out that the defence is built on a fallacy. As was mentioned before, the fact that an event has some probability is compatible with any number of occurrences of that event in an actual sequence of trials.

It is often argued (mistakenly, we think) that an area in which it would not be absurd to act as if some uncertain hypothesis were true is industrial quality control. Even some vigorous opponents of the Neyman-Pearson method, such as A. W. F. Edwards (1972, p. 176), have conceded this. The argument is the following. Suppose an industrialist would lose money if a production run included more than a certain percentage of elements that were defective. And suppose production runs were each sampled with a view to testing whether they were of the loss-making type or not. In such cases there could be no graduated response, it is claimed, since the production run would either be marketed or not. The argument then continues: if the industrialist followed Neyman-Pearson principles, he could comfort himself with the thought that since he will perform the same test and apply the same decision-rule many times, in the long run only about 5 per cent of the batches he will market will be defective, and that may be a failure rate he can sustain. This argument, however, involves a fallacy, for just because the industrialist is confined to one of two actions does not mean he can only either accept or reject the null hypothesis. The situation is compatible with his attaching probabilities to the various hypotheses and then deciding whether to market the batch or withhold it by balancing those probabilities against the utilities of the possible outcomes of his actions. (The precise way in which this is done is the subject of what is known as Decision Theory.) The type of approach to decision making just outlined is, surely, closer to both actual and reasonable calculations. For if an industrialist really accepted some hypothesis about the constitution of a batch as definitely true, he would not only be prepared to sell it on the market, he would also accept any bet whatsoever on the truth of the hypothesis, which he clearly would not. We conclude, therefore, that industrial quality control is not a case that either should or does exemplify Neyman-Pearson thinking.

c.2 The Neyman-Pearson Theory as An Account of Inductive Support

A further reason why it is so difficult to interpret acceptance and rejection as the kind of behaviour one should expect from scientists is that reasonable people do not view theories in such black-and-white terms. As the quotations given in Chapter 1 illustrate, scientific hypotheses are viewed, typically, in varying shades of grey, depending on the weight of evidence in their favour. And it is at least partly the strength of such evidence which dictates how much a scientist will risk on actions whose effectiveness is calculated on the basis of the theory.

This fact is acknowledged by the many classical statisticians who have attempted to interpret their methods of inference in a way which would allow the impact of evidence on a theory to be a matter of degree. The theory of significance tests appears, on the surface, to have room for such an interpretation. In particular, it is commonly assumed that a theory rejected in a statistical test is the more substantially discredited the smaller the size of the test; and if it is accepted, then its merit in the light of the evidence is sometimes measured by the power of the test.

For instance, Kendall and Stuart have suggested that the results of a Neyman-Pearson test can indicate whether or not a hypothesis is supported by the evidence:

> If the reader cannot overcome his philosophical dislike of these admittedly inapposite expressions, he will perhaps agree to regard them as code words, "reject" standing for "decide that the observations are unfavourable to" and "accept" for the opposite. (Kendall and Stuart, 1979, p. 177)

And at the outset of their exposition of the theory of significance tests, Kendall and Stuart stated that its aim is to answer the question: "which sets of observations are we to regard as favouring, and which as disfavouring, a given hypothesis?" (Kendall and Stuart, 1979, p. 177).

A similar view has been expressed by Cramér (1946, p. 421), who wrote: "we shall denote a value [of a test-statistic] exceeding the 5% limit but not the 1% limit as *almost significant,* a value between the 1% and 0.1% limits as *significant,* and a value exceeding the 0.1% limit as *highly significant.*" Cramér added that such terminology is "purely conventional"; nevertheless, it is highly suggestive and seems to have been introduced specifically because of the feeling that a significance test should provide an index of how strongly evidence tells either for or against the hypothesis under test. Thus, on the following page of his book, Cramér remarked that since a particular χ^2 value, calculated from some of Mendel's famous experiments on pea plants, differed by only a small amount from the expected value relative to the null hypothesis he was considering, "the agreement must be regarded as good" (p. 422). In another example, when the hypothesis would only be rejected if the level of significance were around 0.9, Cramér said that "the agreement is very good" (p. 423).

Kendall and Stuart go along with this idea that the size of the test which would just lead to a hypothesis being rejected measures the support which it gives the hypothesis. In an example of theirs, where a χ^2-statistic evaluated from some experiment would reject a hypothesis if the size of the test exceeded 0.37, the agreement between hypothesis and data was said to be "very satisfactory". However, when the critical test-size was only 0.27, Kendall and Stuart (1979, p. 458) described the result as being "still very satisfactory", though they considered it to be "rather more critical of the hypothesis than the other test was". (Oddly enough, these two results were calculated by different methods of grouping the very same data, and there is no indication as to which of the two methods is intended to take precedence. The reader is referred back to section **f** of Chapter 5 for a discussion of this peculiar aspect of the chi-square test.) In general, then, these authors take the view that the smaller the size of the test which would just result in the rejection of the hypothesis, the more telling is the evidence against that hypothesis.

As far as one can tell, however, there are no grounds for

this thesis. In order to establish it, one would need to have formulated an appropriate concept of empirical support, which could then be shown to have the aforementioned connection with the size of significance tests. But this has not been done. Indeed, there are reasons to think that the concepts of size and power are quite unpromising foundations for a theory of empirical support. First of all, the results of a significance test, either of the Fisher or Neyman-Pearson variety, are often in flat contradiction to the conclusions which an impartial scientist or ordinary observer would draw. Secondly, we shall find that judgments based on significance tests are importantly different from decisions about inductive support, for the former depend on what our intuitions in the matter, such as they are, affirm to be extraneous considerations. The rest of this section details these objections.

c.3 A Well-Supported Hypothesis Rejected in a Significance Test

Consider the first objection, which can be explained by referring again to the tulip-bulb example. Let us continue to regard h_1, the hypothesis that the consignment contained 40 per cent red-flowering bulbs, as the null hypothesis. The following table gives the minimum number of red-flowered plants which would have to appear in a random sample of size n in order for that hypothesis to be rejected at the 5 per cent level.

TABLE IX

Sample size n	The minimum number of red tulips needed to reject h_1 at the 5% level, expressed as a proportion of n	Power of the test against h_2
10	0.70	0.37
20	0.60	0.50
50	0.50	0.93
100	0.480	0.99
1000	0.426	1.0
10,000	0.4080	1.0
100,000	0.4026	1.0

It will be noticed that, as n increases, the critical proportion of red tulips that would reject h_1 at the 0.05 level approaches more closely to 40 per cent, that is, to the proportion which h_1 asserts is contained in the consignment. Bearing in mind that the only alternative to h_1 that is admitted in this simple example is that the consignment contains red tulips in the proportion of 60 per cent, an unprejudiced consideration of these data, would, it seems to us, lead one to conclude that as n increases, these so-called critical values *support* h_1 more and more. When the sample size is 1000, the results of the experiment, far from being unfavourable (in Kendall and Stuart's phrase) to h_1, appear strongly to favour it. Yet the theory of significance tests requires one to reject that hypothesis, an injunction that is markedly counter-intuitive.

The example envisages that h_1 is 'tested against' h_2, that is, h_2 is regarded as the only alternative to h_1. In these circumstances it is a simple matter to calculate the power of the test we are employing and this information is included in Table IX. The thesis implicit in the current approach, that a hypothesis may be rejected with increasing confidence or reasonableness as the power of the test increases, is not borne out in the example, which signals the reverse trend. (This objection was developed by Lindley, 1957.)

c.4 A Subjective Element in Neyman-Pearson Testing: The Choice of Null Hypothesis

Another reason why significance tests cannot produce an account of inductive support is that the results of such tests hang on decisions which would normally be regarded as quite unconnected with the evidential force of experimental data. The first kind of decision concerns the selection of the null hypothesis, and it has a crucial bearing on which hypothesis is finally accepted and which rejected, and hence on which is supported and which undermined by the evidence. Thus in the tulip-bulb example, if an experiment produced 50 red tulips in a random sample of 100, then, with h_1 (40 per cent red) as the null hypothesis, it would be rejected at the 5 per cent level and h_2 (60 per cent red) would be accepted. If, on the other hand, h_2 were the null hypothesis, then the opposite judgment would be delivered, that is, h_2 should be rejected and h_1 accepted! But as we observed earlier, the null hypothesis is just the one which according to the scientist's personal scale of values would lead to the more undesirable practical consequences, if it were mistakenly assumed to be true. When the scientist is unable to distinguish the hypotheses by this practical criterion, the null hypothesis may be chosen arbitrarily. But our ordinary understanding of empirical support, such as it is, accords no role to such factors.

c.5 A Further Subjective Element: Determining the Outcome Space

In any kind of significance test, a hypothesis is assessed according to the probability of the observed outcome relative to that of other possible outcomes. In other words, whether an observation favours a hypothesis or not depends upon what observations might have been made if the trial had not turned out as it did. This means the significance of a result of, say, 1 head and 3 tails in an ordinary coin-tossing trial would depend on how the experimental apparatus dealt with other possible outcomes. The apparatus might, for example, be rigged up only to record two results: either the occurrence of 1 head and 3 tails, or else its non-occurrence. In fact, the apparatus might have been designed to report the non-occurring results in many different ways. Each would demand a separate analysis, with no guarantee of always delivering the same significance-test inference.

The outcome space of a trial and hence the decision whether to reject a hypothesis or not is also governed by the so-called *stopping rule,* the rule that dictates when the trial should terminate. Thus, consider again the test of the fair-coin hypothesis with which we illustrated Fisher's theory. We assumed there a stopping rule with the instruction to conclude the trial after 20 flips of the coin. This implies an outcome space of 2^{20} sequences, each comprising 20 elements, where each element corresponds either to heads or tails.

Suppose, however, the rule had been to stop the experiment after 6 heads appeared. The outcome space is now different from that just described, and many of the results that were possible when the number of tosses of the coin was pre-set at 20 are not any longer possible. If, as before, one ignored the order in which heads and tails appear in a trial and simply noted their number, then the results one could record are (6,0), (6,1), (6,2), . . . , etc., whereas before they were (20,0), (19,1), . . . , (0,20)—see Table I, Chapter 5. This change, the reader may be surprised to learn, has a profound effect on the conclusion of a significance test.

This can be appreciated by calculating the probability of each of the possible results under the new stopping rule, relative to the null hypothesis, namely that the coin is fair. The calculation proceeds by considering that the result $(6,i)$ is obtained when $(5,i)$ appears in any order, and is then followed by a head. So, on the assumption of the null hypothesis, the probability of $(6,i)$ is $^{i+5}C_5(\frac{1}{2})^5(\frac{1}{2})^i \times \frac{1}{2}$, yielding the following distribution:

TABLE X

Outcome (H,T)	Probability	Outcome (H,T)	Probability
6,0	0.0156	6,11	0.0333
6,1	0.0469	6,12	0.0236
6,2	0.0820	6,13	0.0163
6,3	0.1094	6,14	0.0111
6,4	0.1230	6,15	0.0074
6,5	0.1230	6,16	0.0048
6,6	0.1128	6,17	0.0031
6,7	0.0967	6,18	0.0020
6,8	0.0786	6,19	0.0013
6,9	0.0611	6,20	0.0008
6,10	0.0458	6,21	0.0005
			etc.

Suppose 6 heads and 14 tails were obtained in a trial conducted according to the present stopping rule. The outcomes which could have occurred and which are at least as improbable as the actual one are $(6,14)$, $(6,15)$, ... Since the total probability of these is only 0.0319, which is less than the critical value of 0.05, the result of the trial is significant and so, according to Fisher, the null hypothesis should be rejected. It will be recalled, with concern we hope, that when the stopping rule restricted the size of a sample to 20, this result was not significant, according to Fisher. (An example similar to this is given by Lindley and Phillips, 1976.) The same contradictory conclusions would be permitted by a Neyman-Pearson treatment if the null hypothesis were tested against some alternative hypothesis, for example, one asserting that the probability of tails in a toss of the coin is P', where P' is greater than 0.5.

Thus a result that was not significant when arrived at with one stopping rule in mind is so when another was intended.

This dependence of inferences upon the stopping rule is a general feature of the classical approach. We regard it as highly damaging to the idea that a significance test can supply any measure of empirical support. (The same point is made by Savage, 1962, p. 18.)

Some stopping rules demand a more complex analysis than that given above. For example, if the experimenter had decided to end the trial when his wife called him for lunch, one would have to evaluate for every stage of the trial the probability of the scientist's being interrupted by his wife's summons. If this could not be done, then the experiment would have been useless. The particular experimental result could even be the product of many, conflicting stopping rules. For a scientist might have acted on one such rule, while his collaborator had privately decided to conduct the trial in accordance with another, and yet the outcome of the trial might have happened to satisfy both rules. There may or may not be an outcome space in such cases, depending on what each experimenter would have done if the actual outcome had not accorded with his own stopping rule. Would he have tried to overrule his colleague or given in quietly? And if the former, would he have succeeded? Intuitively, such questions do not seem to bear on the testing process, and in practice they are rarely if ever put, yet for the classical statistician, they are vital.

A significance test inference, therefore, depends not only on the outcome that a trial produced, but also on the outcomes that it could have produced but did not. And the latter are determined by certain private intentions of the experimenter, embodying his stopping rule. It seems to us that this fact precludes a significance test delivering any kind of judgment about empirical support, unless our intuitions in the matter of such support were sharply revised. For scientists would not normally regard such personal intentions as proper influences on the support which data give to a hypothesis. In Chapter 10, we shall show that as far as the Bayesian theory of inductive reasoning is concerned, the outcome space and the scientist's intentions that help to create it are irrelevant.

■ d TESTING COMPOSITE HYPOTHESES

We have, hitherto, presented only a truncated version of the Neyman-Pearson method, restricting it to the case of two alternative hypotheses, one of which is assumed to be true and the other false. Since such cases are atypical (according to Hays, 1969, p. 275, in psychological research they are "almost nonexistent"), Neyman and Pearson extended and substantially modified their approach to enable it to embrace instances where a larger number of alternatives are admitted. For example, they provided a method, which we shall now describe, whereby the hypothesis that some population parameter θ has a particular value, say θ_0, (call this h_1) may be "tested against" the composite hypothesis $\theta > \theta_0$ (h_2).

Imagine a trial in which a random sample is drawn from the population and appropriately measured. The old approach can still be used to establish a critical region corresponding to any designated probability of committing a type I error. However, since h_2 is equivalent to a disjunction of hypotheses, the calculation of the probability of a type II error is not so straightforward; the probability depends on which element of h_2 is true and this is unknown. h_2 itself assigns no probability to events in the outcome space; hence, there is no determinable probability of a type II error in tests of composite hypotheses. This does not imply that such errors may not occur, of course. It does mean, however, that the criterion of low size and high power cannot be applied to the more complicated situation.

The response to this has been to advance a new criterion, which may be explained as follows. Consider, first, the separate hypotheses disjoined in h_2 and calculate, on the basis of some fixed significance level, the probability of a type II error for each. The critical region that minimizes this probability for every component of h_2 is said to be a *uniformly most powerful* (or UMP) test. In other words, an UMP test is a test of maximum power for the given significance level, whatever hypothesis encompassed by h_2 happens to be true. This is a strong recommendation within the Neyman-Pearson scheme.

A weakness of UMP tests, however, is that relative to some elements of h_2, including perhaps the true one, the power may be very low. Indeed, it could happen that the power is smaller than the significance level, which would mean that there was a greater chance of rejecting the null hypothesis when it is true than when it is false. If this were the case, there would always be a test which, in Neyman-Pearson terms, is better. Moreover, this test need not even involve sampling the population; one could simply sample a pack of cards. For example, if the pack consisted of 5 red and 95 black cards and the decision whether to accept or reject the hypothesis were determined respectively by whether a black or a red card was drawn in a random selection from the pack, the chances of rejecting the hypothesis would be the same whether it was true or false; in this case it would be 0.05. The idea that one could perform a satisfactory test simply by consulting a pack of cards, without even examining the population referred to in the hypotheses being tested, is rightly regarded by classical statisticians as absurd. They hold this view even though randomized trials (see the end of section **a,** above) appeal to the very similar idea that a randomly selected card, or some other extraneous random event, can influence a significance-test inference. Be that as it may, Neyman and Pearson imposed the further restriction on UMP tests that they should be unbiased.

A test of h_1 against h_2 is said to be *unbiased* if its power is at least as great as its significance level, in other words, if there is not a greater chance of rejecting h_1 when it is true than when it is false. When h_2 is composite, the test is unbiased if this condition applies to every element of h_2. Neyman and Pearson then suggested that one ought to confine UMP tests to those that are also unbiased, or to UMPU tests. These ideas can be illustrated by means of a simple example. Consider a population that is normally distributed, with a known standard deviation but an unknown mean, which we can label θ, and a test of the null hypothesis $\theta = \theta_0$ against the composite hypothesis $\theta > \theta_0$. Let the test be based on the mean, $\bar{x}$, of a random sample drawn from the population. The following represents the probability distributions of such samples relative to h_1 and to some hypothesis, h_i, that is included in h_2.

AN UMPU TEST

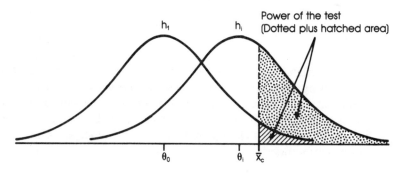

Consider a critical region for rejecting h_1 consisting of points to the right of the critical value, $\bar{x}_c$. The area of the hatched portion is proportional to the significance level. The area under the h_i curve to the right of $\bar{x}_c$ represents the probability of rejecting h_1 when it is false, which is the power of the test. Clearly, if the standard deviation remained unchanged, the closer θ_i was to θ_0, the smaller the power. But however close θ_i approached θ_0, the power could never fall below the level of significance. In other words, the critical region is an UMPU test for h_1 against h_2. However, although the test is uniformly most powerful and unbiased, its actual power is unknown and, as already pointed out, it might be only infinitesimally different from the significance level. In other words, although the test is unbiased, it may be only just unbiased.

Suppose, now, the case is one in which h_1: $\theta = \theta_0$ is tested against h_2: $\theta \neq \theta_0$, it again being assumed that the standard deviation is known. A critical region located in one tail of the h_1 curve, as before, would not now constitute an unbiased test, as can be seen by considering a hypothesis h_i, in h_2, that is centred around the other tail, a situation depicted in the next diagram.

A BIASED TEST

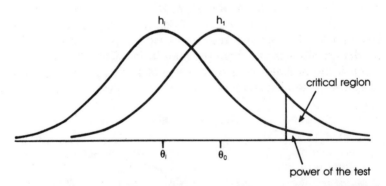

An UMPU test can, however, be constructed for this case by dividing the critical region equally between the two tails. This can be appreciated diagrammatically and can be rigorously shown. An UMPU test thus reinstates the two-tailed test implied by Fisher's theory, but for which Fisher gave no adequate rationale.

In order to deal with composite hypotheses, the Neyman-Pearson method needs to depart somewhat from the original ideal. Thus, although the principle of maximizing power for a given level of significance is preserved, one cannot discover what that power is, and, for all one knows, in any particular case the power may be very low. This seems to be a disadvantage, considering that the power of a test was, as Hays (1969, p. 270) put it, intended to reflect "the ability of a decision-rule to detect from evidence that the true situation differs from a hypothetical one". UMPU tests represent a more significant dilution of the original ideal, for they do not have maximum power. It is also worth mentioning that whatever interest UMP and UMPU tests may have is necessarily rather academic since there are so few circumstances in which they exist. Far more serious, however, than any objection that might be mounted along these lines is the fact that the modifications introduced to deal with composite hypotheses are equally afflicted by the various difficulties that we have already shown to discredit the Neyman-Pearson method in the simple case.

■ e CONCLUSION

The significance test in its various forms was created to supplant Bayes's Theorem as an instrument of inference, in the hope that statistical inductive reasoning could thereby be placed on an entirely objective and rational footing. We have argued that neither of these ideals is actually served. Avoiding the probabilistic conclusion that a Bayesian analysis affords turns out to be rather difficult, and the various possible non-probabilistic interpretations that might be placed on the conclusion of a significance test all prove unsatisfactory. Inference by significance test also clashes with entrenched ideas about the nature of evidence, requiring the rejection of hypotheses that seem highly confirmed, allowing (in randomized tests) quite extraneous experiments such as the selection of cards from a pack to influence one's attitude towards hypotheses which have nothing to do with cards, and investing the stopping rule with an importance which intuitively it does not possess. It does these things without supplying any good reason why one's intuitions should be overridden. The role of the stopping rule also highlights the fact that a test of significance must treat as real that which is purely imaginary and which may vary from one imagination to another, namely, the outcome space.

It is ironic as well as instructive that so considerable an effort to produce a purely objective method should give birth to something so different from that originally conceived. First, there are the personal decisions about what level of significance to adopt and which hypothesis should be designated the null hypothesis. In Fisher's approach, there is the decision about which test-statistic to employ. One must also have a means of limiting the hypotheses to which the test is applied, whereby serious contenders are separated from other hypotheses, and this, of course, cannot be done through classical statisticians' tests of significance. And Fisher's account of how causal hypotheses should be tested requires a personal decision on which factors in the experimental situation should be randomized. Pearson regarded such subjective elements as inevitable:

> Of necessity, as it seemed to us [him and Neyman], we left in our mathematical model a gap for the exercise of a more intuitive process of personal judgement in such matters—to use our terminology—as the choice of the most likely class of admissible hypotheses, the appropriate significance level, the magnitude of worthwhile effects and the balance of utilities. (Pearson, 1966, p. 277)

In the next chapter we shall see that attempts to extend classical, non-probabilistic, inference to other areas makes the inference process even more dependent on personal judgments. When thinking of the antagonism of the founders of the classical outlook to the Bayesian approach on the grounds of its subjectivity, one is inevitably reminded of those who would strain at a gnat while contentedly swallowing a camel.

CHAPTER 8

The Classical Theory of Estimation

■ a INTRODUCTION

Thus far we have considered inferences, based on significance tests, in which a decision is taken on whether or not "some pre-designated value [of a parameter] is acceptable in the light of the observations" (Kendall and Stuart, 1979, p. 175). Many classical statisticians find such conclusions inadequate for scientific purposes, which, they believe, require "principles upon which observational data may be used to estimate, or throw light upon the values of theoretical quantities, not known numerically" (Fisher, 1956, p. 140). Or, as Kendall and Stuart put it, "We require to determine, with the aid of observations, a number which can be taken to be the value of [some unknown parameter] θ, or a range of numbers which can be taken to include that value" (1979, p. 1). In other words, something more precise is wanted than a fallible acceptance that θ exceeds some value or is unequal to a specific number. The classical theory of estimation, created very largely by Fisher and Neyman, was designed to meet this need.

Scientists frequently estimate the value of a physical quantity and in the process come to regard a certain number as a good approximation to the true value. For instance, a chemist would estimate the boiling point of a liquid by averaging several carefully made thermometric readings. And the average height of people in a large population would be estimated by the mean of an appropriate sample. It is natural to adopt a more or less tentative attitude towards estimates arrived at in these ways and to view their accuracy as improving with the size and representativeness of the samples and with the perceived reliability of the measuring devices.

All this is explicable in Bayesian terms. Before the experiment, the scientist typically has only a rough idea of a parameter's value, and so his beliefs would be described by a rather diffuse prior probability distribution. That is, a relatively wide range of possible values would be roughly similar in probability, and no single value or (for the scientist's purposes) sufficiently narrow band of values would stand out as being very probable. But evidence from a well-designed experiment may induce a posterior probability distribution that is more concentrated around a particular value, the relation between the posterior and prior distributions being expressed in Bayes's Theorem. In most cases, the larger and more representative the sample, the more the posterior probability crowds around some particular value; hence, the greater is the chance that the true parameter value lies in that neighbourhood. (A more detailed account of the principles governing this process will be found in Chapter 10.)

Classical statisticians, on the other hand, have looked for principles of estimation which do not assume the unknown parameter to be a random variable and which do not rely on Bayes's Theorem. In a sense they have been successful, for classical estimation procedures have attained a high level of sophistication and now exert an enormous influence on statistical workers and scientists concerned with statistical hypotheses. However, in our view, which we shall seek to substantiate, this influence is undeserved. We shall, as in our consideration of significance tests, enquire whether the classical principles of estimation are reasonable, and we shall also investigate whether they manage to avoid the kind of subjectivity which its founders and many practitioners found objectionable in the Bayesian system.

Classical estimation theory has two branches, known as *point estimation* and *interval estimation*. Point estimation aims to select a specific number as the so-called best estimate of a parameter; it is contrasted in the literature with interval estimation, a method of locating the parameter within a region and associating a certain degree of 'confidence' with the conclusion that is drawn. These two approaches employ difference techniques, and we shall follow standard expositions by dealing with them separately.

■ b POINT ESTIMATION

Point estimation is intended to solve the problem posed "when we are interested in some numerical characteristic of an unknown distribution (such as the mean or variance in the case of a distribution on the line) and we wish to calculate, from an observation, a number which, we infer, is an approximation to the numerical characteristic in question". (Silvey, 1970, p. 18) One could, of course, pick any arbitrary number and offer it as an estimate of a physical quantity, but this would not be a reliable technique in general, and in any particular case one would have no idea how accurate it was. Statisticians, on the other hand, would like "an estimate based on the sample [which is] generally better than a sheer guess". (Lindgren, 1976, p. 253) Of course the classical theory does not suppose that estimates arrived at from samples are infallible. On the other hand, it denies that one can meaningfully assign a real, or objective, probability to the proposition that a parameter's true value equals, or is close to, its estimated value; and classical statisticians repudiate the idea that scientists should rely on subjective probability appraisals of theories.

The classical technique for estimating any population parameter is the following. First, a sample is drawn from the population and each element of the sample measured. Suppose $\mathbf{x} = x_1, \ldots, x_n$ denotes the measurements thus derived in a sample of size n. An estimating statistic, t, is selected, this taking the form of a calculable function $t = f(\mathbf{x})$. Finally, the inference is drawn that t_o, the value of t derived from a particular experiment, is the best estimate of the unknown parameter. As classical statisticians deny that such an estimate can be qualified by the probability of its being accurate, they have invented methods by which to decide its quality indirectly. An estimate is said to be 'good' if the method which produced it is 'good'. So-called good methods of estimation, or good estimators, are ones that satisfy certain desiderata, the most frequently mentioned restrictions on methods of estimation being that they should be *sufficient, unbiased, consistent,* and *efficient.* We shall explain these concepts in turn and assess their supposed relevance to estimation.

b.1 Sufficient Estimators

We introduced the notion of sufficiency earlier, in the context of Fisherian tests of significance. It will be recalled that sufficiency in a statistic is usually interpreted as signifying that the statistic contains all the relevant information about some parameter, θ. This is a plausible interpretation, for a statistic

t is defined to be *sufficient* for θ when the probability $P(\mathbf{x} \mid t)$ is the same for all θ, where $\mathbf{x} = x_1, \ldots, x_n$ is any element of the outcome space. In other words, once the value of t has been fixed in an experiment, more specific details of the outcome will not depend on θ. On these grounds, sufficiency is standardly incorporated as a criterion for selecting estimators for point estimation. Accordingly, classical statisticians endorse the sample mean, which unlike the sample median or range, say, is sufficient for estimating the population mean.

Bayesians also regard sufficient statistics as containing all the information that is needed to estimate a parameter. But while this must just be incorporated as a desideratum into classical point estimation theory, the Bayesian account has a proof for it. The Bayesian interpretation of the classical condition is that the statistic $t = f(\mathbf{x})$ is sufficient for the parameter θ just in case $P(\mathbf{x} \mid t \ \& \ \theta) = P(\mathbf{x} \mid t)$. This condition would, of course, not be regarded as meaningful in the classical scheme, for it relies on probability terms conditioned upon θ and thus treats the parameter as a random variable, which, classically speaking, is impermissible, since it is not itself the outcome of a repeatable experiment over which a physical probability distribution exists. However, the condition expresses in Bayesian terms the same idea which classical statisticians hoped the condition of sufficiency would capture in classical terms. Both conditions assert that t is sufficient for θ just in case $\mathbf{x}$ is independent of θ, once t is given; in the former case, this independence is a functional independence, in the latter, it is probabilistic.

From the Bayesian point of view, sufficient statistics sacrifice no relevant information, because the conclusion of a Bayesian inference, describing a posterior probability, is exactly the same whether one uses the raw data or a sufficient statistic based on those data. In other words, $P(\theta \mid t) = P(\theta \mid \mathbf{x})$. The proof is as follows: according to Bayes's Theorem, for all values of θ, $\mathbf{x}$, and t

$$P(\mathbf{x} \mid \theta \ \& \ t) = \frac{P(\theta \mid \mathbf{x} \ \& \ t)P(\mathbf{x} \mid t)}{P(\theta \mid t)} \ .$$

In the continuous case, the corresponding form of the theorem employing densities must be substituted, the following argument proceeding mutatis mutandis. Since $\mathbf{x}$ uniquely determines t, so that $\mathbf{x} \ \& \ t$ is logically equivalent to $\mathbf{x}$, it follows that

$P(\theta \mid \mathbf{x} \& t) = P(\theta \mid \mathbf{x})$. Combining this with Bayes's Theorem gives

$$P(\mathbf{x} \mid \theta \& t) = \frac{P(\theta \mid \mathbf{x})P(\mathbf{x} \mid t)}{P(\theta \mid t)} \ .$$

This implies that $P(\theta \mid \mathbf{x}) = P(\theta \mid t)$, just in case the condition for sufficiency, $P(\mathbf{x} \mid t \& \theta) = P(\mathbf{x} \mid t)$, holds. In fact, Bayesians often use the former rather than the latter equation as the defining characteristic of sufficiency.

The total evidence requirement. Classical statisticians take it for granted that an estimate, or the conclusion of a significance test, should use all the relevant information in a sample, as is more or less self-evident that one should. But despite the naturalness of the requirement, it has to be asserted as a separate postulate in both classical and Bayesian methodologies, the postulate usually being known as the Total Evidence Requirement (Carnap, 1947). Once this is conceded, it follows that a Bayesian analysis must employ sufficient statistics.

This argument proceeds from a theory of evidential or inductive relevance (that is, the Bayesian theory) to a demonstration that, as intuition affirms, only sufficient statistics contain all the information that is relevant for estimating a parameter. From the Bayesian viewpoint, therefore, estimation must be conducted with sufficient statistics. No analogous argument has been constructed from the classical theory of point estimation. Defenders of that theory seem forced to argue in a reverse order from the Bayesian; that is, they must start from the intuition that only sufficient statistics include all the relevant information, and then incorporate sufficiency as a criterion for estimators. However, they have provided no evidence that the intuition is well-founded nor any sound explanation for the source of the intuition. In our opinion the source is none other than Bayes's Theorem, applied unconsciously.

b.2 Unbiased Estimators
The notion of an unbiased estimator (which is different from that of an unbiased test discussed in the last chapter) is defined in terms of the expectation, or expected value, of a random variable. The latter, as already mentioned, is given by the expression: $E(x) = \Sigma x_i P(x_i)$, the sum (or, in the continuous case, the integral) being taken over all the values that the

variable x can assume. The expectation of the random variable x is also called the mean of the probability (or density) distribution of x. When x is symmetrically distributed (according to the normal law, for example) the mean or expected value of x is also the geometrical centre of the distribution. In such cases, the expected value of x is also the most probable value.

A statistic t is defined to be an *unbiased estimator* of a parameter if its expectation is equal to the parameter's true value. If t is unbiased in this sense, then, other things being equal, the value of t derived from a particular experiment is said to be a good estimate of the parameter.

Many estimators one would favour on intuitive grounds are, in fact, unbiased. For instance, the proportion of red counters in a sample is an unbiased estimator of the corresponding proportion in the urn from which they were randomly drawn. Similarly, a sample mean is an unbiased estimator of the corresponding population mean. However, although theory and practice meet in many such cases, we must consider whether this might be merely fortuitous or whether unbiasedness is necessary for estimators.

The term that classical statisticians chose for the criterion we are discussing intimates that bias is in some way bad for an estimator, and it was clearly introduced for its suggestion of fair-mindedness and lack of prejudice. This impression is reinforced by the fact that biased estimators are standardly 'corrected'; indeed, general methods have been invented for making such 'corrections for bias'. Thus Kendall and Stuart inform us that while intuition may suggest the sample variance as an estimator for the variance of the parent population, "that intuition is not a very reliable guide in such matters" (1979, p. 4). The alleged difficulty is that the sample variance is a biased estimator of the corresponding population variance. On the other hand, the corrected statistic obtained by multiplying the sample variance by the factor $\dfrac{n}{(n-1)}$ is unbiased "and for this reason it is usually preferred [as an estimator] to the sample variance". (Kendall and Stuart, 1979, p. 5) But, these bold claims and recommendations on behalf of unbiased estimators are not easily sustained.

Consider how the classical statistician recommends we proceed. In estimating a parameter, one must select an unbiased estimator t. Suppose a particular experiment gives the result

$t = t_o$. We are now invited, other things being equal, to infer that the true or approximate value of θ is t_o, or, as Kendall and Stuart have expressed it, that t_o "can be taken to be the value of θ".

But there is no necessary connection between an estimator's unbiasedness and the propriety of reposing confidence in its specific estimates; any particular unbiased estimate may, for all we know, be very inaccurate, and we are certainly not entitled to infer anything about its probability of being close to the truth. It is therefore not surprising that statisticians' recommendations of the unbiasedness criterion are sometimes rather muted. For example, Barnett (1973, p. 120) said that "within the classical approach unbiasedness is often introduced as a *practical* requirement to limit the class of estimators within which an optimum one is being sought" (our italics). Even Kendall and Stuart, who have written so confidently of the need to correct estimators if they are biased, conceded that the prominent place accorded to unbiasedness in classical estimation theory has no epistemic underpinning. Astonishingly, they admitted that

> There is *nothing except convenience* to exalt the arithmetic mean [i.e., the expectation] above other measures of location as a criterion of bias. We might *equally well* have chosen the median of the distribution of [an estimator] *t* or its mode as determining the "unbiased" estimator. The mean value is used, as always, for its mathematical convenience. (Kendall and Stuart, 1979, p. 4; our italics)

By this Kendall and Stuart are saying that one might just as well have defined an estimator as unbiased when its median or its mode, instead of its mean, is equal to the population parameter. But of course each of these alternatives is compatible with a different class of estimators, and there is no guarantee whatever that in particular cases actual estimates would be identical or even similar. So an unbiased estimate can be called the best only in a limited sense; it is not the best in virtue of being the most accurate, or the most probable, or the most rationally believable, but simply because it was derived with a minimum of tiresome mathematical computation.

Indeed, Kendall and Stuart (1979, p. 4) confirmed that the bias-criterion lacks epistemic force when they warned readers that "the term 'unbiased' should not be allowed to convey overtones of a non-technical nature". Such overtones are, of course, inevitably conveyed, and the originators of the idea that unbiasedness should serve as a standard for good estimation no doubt selected this suggestive name deliberately. Unfortunately, expositions of the classical point of view usually ignore the purely technical meaning of bias. If a more neutral name had been chosen, this oversight would surely be encountered less frequently, and the criterion of bias would lose what plausibility it possesses and be relinquished as a standard against which our intuitions are judged correct or incorrect. The same might be said of the criterion to be discussed next.

b.3 Consistent Estimators

The conditions of bias and sufficiency do not, by themselves, determine a unique estimator. Hence the need for a number of further criteria. Amongst these is the criterion of consistency. An estimator is said to be *consistent* when its probability distribution shows a diminishing scatter about the true value as the sample size increases. More precisely, a statistic t derived from a sample of size n, is a consistent estimator for θ if, for any arbitrary positive number ϵ, $P(|t - \theta| \leq \epsilon)$ tends to 1, as n tends to infinity. This condition is sometimes referred to as t tending probabilistically to θ. A consistent estimator is said by Kendall and Stuart to exhibit "increasing accuracy", which property they describe as "evidently a very desirable one" (1979, p. 3).

According to Fisher, consistency is the "fundamental criterion of estimation" (1956, p. 141). Indeed, he believed that statistics that were not consistent "should be regarded as outside the pale of decent usage" (1970, p. 11). And Neyman agreed "perfectly" in this judgment, adding that "When one intends to estimate a parameter . . . , it is definitely not profitable to use an inconsistent estimate". (1952, p. 188)

Fisher defended his emphatic preference for consistency with the following somewhat obscure and, in our view, unpersuasive argument:

> . . . as the samples are made larger without limit, the [estimating] statistic will usually tend to some fixed value characteristic of the population, and, therefore, expressible in terms of the parameters of the population. If, therefore, such a statistic is to be used to estimate these parameters, there is only one parametric function to which it can properly be equated. If it be

equated to some other parametric function, we shall be using a statistic which even from an infinite sample does not give a correct value.... (1970, p. 11)

This was meant to establish that consistent estimators and no others should be employed. But its claim is only that when (*not* 'only when') such estimators converge to a particular population parameter can they 'properly be equated' with that parameter alone and not with any other. So Fisher's argument could not show the necessity of the consistency condition. Moreover, Fisher seems to be wrong to claim that an estimate and the quantity being evaluated can ever be properly equated, for, as is granted on all sides, even in the most favourable circumstances, an estimate may be wrong.

Carefully formulated arguments for the consistency requirement do not seem to exist. Its necessity is normally just asserted as more or less obvious, and the plausibility of the assertion is, no doubt, due in part to a name which intimates unassailable virtue. However, consistent estimators are sometimes recommended on the assumption that they become more 'accurate' as the sample size increases, it being presumed (*see* section **b.4**, below) that an estimator's accuracy is connected with the spread of its probability distribution about the true parameter value. But however the notion of accuracy is glossed, the mere fact that an estimator would have been more accurate if a larger sample had been drawn implies nothing at all about its accuracy in actual cases, which surely is what one ought to be interested in.

We may illustrate this point with an example in which an 'inconsistent' method of estimation yields a perfectly satisfactory and confidence-inspiring estimate. Let the goal of the estimation be the mean of some population, and imagine a scientist eccentrically selecting $\bar{x} + (n - 100)\bar{x}^2$ as the estimating statistic, where as before, $\bar{x}$ and n are the sample mean and sample size, respectively. Clearly this odd statistic is not consistent (in the technical sense), for it diverges ever more sharply from the population mean as the sample is enlarged. Nevertheless, for the special case where $n = 100$, the statistic is just the familiar sample mean, which on intuitive grounds gives a perfectly satisfactory estimate. This example illustrates the essential weakness of the classical principle that an estimate must be evaluated relative to the *method* by which it was derived.

A corollary of this principle is that an estimate's worth

depends on who derived it. For suppose statistician A employed the sample mean to estimate a population mean, while B used some non-consistent and/or biased function of the sample mean; and imagine that they each arrived at identical estimates from the same sample. According to classical ideas, since these identical estimates have different pedigrees, they have to be differently evaluated: one would be 'good', the other 'bad'! This, of course, contradicts the difficult-to-gainsay assumption that logically equivalent statements are equally 'good', an assumption enshrined in the probabilistic approach. It also violates the now somewhat battered objectivity ideal that supposedly guides the classical approach.

Hence, we see no merit in consistency as a desideratum, even though many intuitively satisfactory estimators are in fact consistent.

b.4 Efficient Estimators

It is intuitively obvious that the probability distribution of an estimator should have as narrow a spread about the true value of the quantity being estimated as possible. Hence a thermometer reading would command more confidence if made by a sober professional than by an inebriate layman, for the likely spread of results in the former case is smaller than in the latter. These preferences are understandable if referred to a Bayesian framework. For a Bayesian estimate is expressed as a posterior distribution over a set of theories; in the examples we have been dealing with, the theories in question would be different possible values of a parameter. As we shall see in Chapter 10, the smaller the variance of an estimator, the smaller the variance of the resulting posterior distribution and, hence, the narrower the range of values within which the parameter can, with high probability, be located.

Classical statisticians also try to connect the variance of an estimator with the quality of the estimate. First they define one estimator to be the *more efficient* when its probability distribution about the true value has the smaller variance. In a natural extension of this idea, one may compare the variances of a statistic calculated from samples of different sizes. The variances of many statistics are inversely related to sample size, in which case their efficiency would improve with a larger sample. Classical statisticians maintain that provided the other conditions we have mentioned are met, the more efficient an estimator, the better. For instance, Fisher held that the less efficient of two statistics was "definitely inferior . . . in its accuracy" (1970, p. 12). Fisher seems to be saying that the more efficient an estimator, the closer must the estimate it gives be to the truth. But he could not really have meant this. For clearly, in the chancy process of estimation, one can never be sure of any degree of accuracy, even with the best estimating statistic. As Neyman (1952, p. 159) observed, in typical instances, "it is more or less hopeless to expect that a point estimate will ever be equal to the true value".

A more characteristic classical defence is that given by Kendall and Stuart (1979, p. 7), who argued that since "An unbiased consistent estimator with a smaller variance will . . . deviate less, on the average, from the true value than one with a larger variance . . . we may reasonably regard it as better". The idea here seems to be the following: suppose e_1^i and e_2^i are the estimates delivered by separate estimators on the ith trial, and suppose θ is the parameter's true value. Then if the first estimator is the more efficient, there is a high probability that $|e_1^i - \theta| < |e_2^i - \theta|$. Kendall and Stuart then translate this high probability into an average frequency in a long run of trials. As we have remarked before, this translation goes beyond logic, though it is quite plausible. Even overlooking this difficulty, the justification of the relative efficiency criterion in terms of average behaviour in a long run of estimations has nothing to say about the accuracy of particular estimates. Since estimates often form the basis of practical actions, and since they are usually expensive and troublesome to obtain, it seems reasonable to demand an assessment of their accuracy.

In practice, this demand is evidently met, for estimates are not normally presented as specific numbers, as a point estimate would be. Thus, one normally finds an estimate stated as a *range* of numbers, in the form $\theta = \theta^* \pm \epsilon$, the suggestion being that it is reasonable to expect this range to contain the true value. A Bayesian would say that θ probably falls in the range. Neyman attempted to give a classical expression to this idea with his theory of interval estimation, which he developed around 1930 and which is now the dominant theory in statistical estimation. We turn next to that theory.

■ c INTERVAL ESTIMATION AND CONFIDENCE INTERVALS

Instead of proposing a specific number for a parameter, as point estimates do, an interval estimate purports to provide "an estimated range of values with a given high probability of covering the true . . . value" (Hays, 1969, p. 288). Estimates of the latter kind are permitted in circumstances where one is able to derive numbers, say t_1 and t_2, from the data, such that if the true value of an unknown parameter is θ, the following equation holds:

$$P(t_2 \leq \theta \leq t_1) = 1 - \alpha.$$

The limits t_1 and t_2 give a so-called *confidence interval* for the parameter, $1 - \alpha$ being its associated *confidence coefficient*. Consider a standard example. A random sample of some specified size is drawn from a population with a mean θ and a finite standard deviation σ. The sampling distribution of means is approximately normal for large samples and its own mean is θ. Its standard deviation, denoted by $\sigma_{\bar{x}}$, is a function of the sample size and of the population's standard deviation (in fact, $\sigma_{\bar{x}} = \sigma n^{-\frac{1}{2}}$), but it is independent of the particular value of θ.

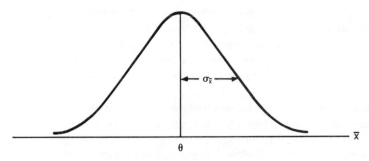

The sampling distribution of means
from a population with mean θ
and standard deviation σ.

If the sample mean is $\bar{x}$, then it follows from the distribution's normality that, with probability 0.95,

$$-1.96 \times \sigma_{\bar{x}} \leq \theta - \bar{x} \leq +1.96 \times \sigma_{\bar{x}}.$$

And this implies that the following has a probability of 0.95:

$$\bar{x} - 1.96 \times \sigma_{\bar{x}} \leq \theta \leq \bar{x} + 1.96 \times \sigma_{\bar{x}}.$$

This range of values is said to be a 95 per cent confidence interval for θ.

c.1 The Inferences Drawn by Classical Statisticians from Confidence Intervals

The technical results concerning confidence intervals, such as that given above, are not in doubt. They "were *deduced* from the specified assumptions regarding the observable random variables and, therefore, are the result of *deductive reasoning*". (Neyman, 1952, p. 209) But the theory of estimation by confidence intervals is inductive, that is, it purports to explain how the properties of a given sample yield information about a wider population. The question arises, therefore, what specific conclusion concerning a parameter value are we entitled to draw on the basis of a confidence interval?

The statement of a confidence interval is easily confused with that of a posterior probability; it seems to state the posterior probability, given the sample, that the parameter lies between the confidence bounds. However, since no corresponding prior probability was stipulated upon which such a posterior probability would necessarily depend and which might take any value from zero to 1, this cannot be the case. The confidence interval does not give a posterior probability. It says that *if* the value of the paramater is θ, then with probability 0.95, $\bar{x} + 1.96 \times \sigma_{\bar{x}}$ and $\bar{x} - 1.96 \times \sigma_{\bar{x}}$ lie on either side of θ. It does not ascribe any probability to θ.

What sort of conclusion may one draw from a confidence interval? The standard answer is that we must "*assert* that [the parameter lies between t_1 and t_2] . . . in each case presented for decision", the justification for such an assertion being that "we shall be right in a proportion $1 - \alpha$ of the cases in the long run". (Kendall and Stuart, 1979, p. 110). This injunction and justification derive from Neyman, who told the "practical statistician" to proceed in three steps: first, he should perform the random experiment; secondly, he should calculate the corresponding confidence interval; and, finally, "he must *state*" that the true value of the parameter lies between the two confidence bounds (1937, p. 263; our italics). And although it is never certain that the parameter really lies within the confidence interval, Neyman thought it reasonable for the statistician to assert that it does, because (when the confidence coefficient is 0.99) "in the long run he will be correct in about 99% of all cases". (1937, p. 263)

Neyman was careful to observe, however, that when making this assertion the statistician is not entitled to conclude or to believe that the assertion is true. Indeed, Neyman held the

idea of inductive reasoning to a conclusion or belief to be contradictory, on the grounds that reasoning denotes "the mental
process leading to knowledge" and this "can only be deductive".
Neyman claimed that induction was not a matter of reasoning,
but of behaviour; in particular, the proper result of an interval
estimate was a decision "to behave as if we actually knew" that
the true value of the parameter was between the confidence
bounds (Neyman, 1941, p. 379).

Summary. Neyman's application of confidence intervals to induction has distinct advantages over point estimation. First,
it enables one to arrive at a range of values in which the parameter is estimated to lie, rather than at some specific value,
and this is more in keeping with scientific practice. Secondly,
different ranges of values are associated with their own confidence coefficients, which suggests—though misleadingly we
shall argue—that one is justified in placing greater reliance on
some estimates than on others, as scientists in fact regularly
do. Thirdly, if the variance of the sampling distribution of the
estimating statistic falls, then the confidence interval calculated for a given confidence level becomes narrower, hence, the
precision of the estimate improves. This appears to be the reasoning behind the relative efficiency requirement for an estimator. In the context of point estimation there seemed to be
no adequate rationale for that criterion.

c.2 Why Confidence Intervals Cannot Be Applied to Inductive Reasoning

I The indefinite-repetitions-argument, again. According to the
orthodox view, the practical conclusion of an interval estimate
is the categorical assertion that the parameter in question has
a value within the corresponding interval. And this inductive
leap is said to be legitimate because a large number of repetitions of the trial would rarely lead the scientist into error.
But, as we have already indicated, we regard such inductions
as unwarranted and the supporting argument as fallacious.
(*See* Chapter 5, section **c,** and Chapter 7, section **c.1.**)

Moreover, no scientist would act as if an uncertain estimate
were definitely correct. There are different degrees of acceptance and rejection, as the writings and utterances of scientists
amply show and as is evidenced in the costs which scientists

are prepared to bear should their estimates turn out to be wrong; the better or more reliable or (in Bayesian terms) more probable the estimate, the less tentative the commitment and the higher the warranted stake against being in error.

Neyman has argued, on the other hand, that it is sometimes reasonable to act as if one's real beliefs were false, or as if what one believed to be false were true, illustrating this with the purchasing of accident insurance before a holiday. In making such a purchase, he said, "we surely act against our firm belief that there will be no accident; otherwise, we would probably stay at home" (1941, p. 380). This seems a perverse analysis of the situation. The reasonable person does not have an absolute conviction that there will be no accident; he believes that there is a larger or smaller chance of an accident, depending upon the type of holiday and on other factors, and he acts in accordance with this belief, for example, by proportioning the amount he is prepared to pay for insurance to the degree of risk he perceives.

ll Estimates and degrees of confidence. The confidence-interval approach does appear to have room for the kind of graduated conclusion that we regard as more realistic. This, at any rate, is a popular view. Thus Harnett (1970, p. 199) maintained that the approach allows one "to assert with a certain degree of confidence that the . . . parameter will fall within . . . [a specified] interval" and he regarded this as overcoming "One of the major weaknesses of a point estimate [namely] . . . that it does not permit the expression of any degree of uncertainty about the estimate". Similarly, Mood (1950, p. 222) claimed that although a confidence coefficient "is not a true probability", it is, nevertheless, "a measure of our confidence in the truth of the statement [that the parameter lies within the experimentally computed confidence interval]". The intuition that prompts such a view would normally be endorsed if inferences were mediated by Bayes's Theorem, as we shall see in Chapter 10. But, in our opinion, the intuition is at odds with the confidence interval approach, despite a superficial appearance to the contrary. This is because a confidence interval depends on various personal judgments and decisions, for instance regarding the stopping rule, which, as we argued earlier in the context of significance tests, do not enter into the normal conception of confirmation and have no influence on a scientist's confidence in hypotheses. The difficulty stems from the fact that these decisions determine the outcome space, in other words, the outcomes that could have occurred if the actual one had not. And, intuitively, these unrealised possibilities are of no evidential account.

Consider the role of the stopping rule in relation to the problem we considered before, namely, that of estimating the mean of a population whose standard deviation has already been established. The usual estimating statistic is the sample mean, and the confidence interval is derived from the sampling distribution of means. Now the latter is calculated on the assumption of a fixed sample-size, so that every element of the outcome space has the same size. But this would not be a valid assumption if the experimenter had merely intended to draw as big a sample as he could manage before lunch, or (as is often the case) just used those members of the population who happened to be available. Each of these represents a different stopping rule, and each might produce different interval estimates. The idea that one's confidence in an estimate should depend upon the stopping rule that was in the mind of the experimenter is absurd, especially since a given result might be the work of several experimenters, each operating with his own stopping rule. The physical characteristics of a chemical compound, for example, are estimated without considering how the number of experiments actually carried out was decided upon. Such information is practically never imparted in learned reports. When interval estimates are calculated on the assumption of a particular outcome space, experimenters rarely, if ever, supply evidence that they had deliberately fixed its composition and size in advance. Indeed, there is often evidence that the number of subjects in such trials was due to a mixture of pragmatic and accidental reasons.

The outcome space and hence the interval estimate are influenced by other extraneous decisions too, for instance, decisions on how one will record outcomes that, in the event, did not occur. This is vividly described by Pratt.

> An engineer draws a random sample of electron tubes and measures the plate voltages under certain conditions with a very accurate voltmeter, accurate enough so that measurement error is negligible compared with the variability of the tubes. A statistician examines the measurements, which look normally distributed and vary from 75 to 99 volts with a mean

of 87 and a standard deviation of 4. He makes the ordinary normal analysis, giving a confidence interval for the true mean. Later he visits the engineer's laboratory, and notices that the voltmeter used reads only as far as 100, so the population appears to be 'censored'. This necessitates a new analysis, if the statistician is orthodox. However, the engineer says he has another meter, equally accurate and reading to 1000 volts, which he would have used if any voltage had been over 100. This is a relief to the orthodox statistician, because it means the population was effectively uncensored after all. But the next day the engineer telephones and says, 'I just discovered my high-range voltmeter was not working the day I did the experiment you analysed for me'. The statistician ascertains that the engineer would not have held up the experiment until the meter was fixed, and informs him that a new analysis will be required. The engineer is astounded. He says, 'But the experiment turned out just the same as if the high-range meter had been working. I obtained the precise voltages of my sample anyway, so I learned exactly what I would have learned if the high-range meter had been available. Next you'll be asking about my oscilloscope'. (Pratt, 1962, pp. 314–315)

The engineer is astounded, of course, by the suggestion that the various revelations that so impressed the statistician should disturb his own confidence in the initial estimate. The engineer's response would surely be that of any reasonable scientist. One would require very persuasive reasons to override the natural and traditional attitudes of experienced and successful scientists, reasons which are not forthcoming from the classical theory of confidence intervals.

iii Prior knowledge. The classical methods of estimation, and also of significance testing, are unable to take account of certain relevant information that is available prior to the experiment (*see* Schlaifer, 1959, pp. 665–666). This is not so when the information is 'objective'. Thus, classical statisticians concede that when a parameter has a known, objective prior probability distribution, Bayes's Theorem ought to replace tests of significance and confidence interval estimation. A case in point would be where an urn is selected from a set of half-a-dozen urns, each containing known proportions of red and white counters, the selection being made on the outcome of the throw of a balanced die. The prior probability that any particular urn has been selected is, then, $\frac{1}{6}$. The corresponding posterior probabilities may then be determined by drawing counters at random from the selected urn.

Examples such as these cause no problem for classical inference; however, they are rare. The more difficult case is where one has some prior knowledge but where that knowledge is not of objective probabilities derived from the kind of stochastic process just described. For instance, it is unlikely that the average height of students attending the University of London varies much from college to college. But suppose that on the basis of random samples from two of the colleges, one obtained 95 per cent confidence intervals of $4'0'' \pm 2''$ in one case, and $5'6'' \pm 3''$ in the other. Such a large intercollegiate divergence is clearly unlikely; the normal response would be to say that the first estimate was probably due to a very unrepresentative sample. The idea that one should accept (in whatever sense) that students at an ordinary London college are so remarkably short is absurd. It would be particularly absurd if one were practically certain that the estimate was false, as one would be if exhaustive records of the students' heights had been lost, it simply being recalled that the average was above 5 feet.

The classical response to this could take one of two forms, neither adequate, we believe. It could claim that classical estimation is applicable only when one has no relevant information regarding the parameter. This, however, would impose a considerable restriction upon estimation, for there are very few cases, if any, where one can be said to know nothing at all about a parameter's value. Moreover, although a little knowledge is certainly a dangerous thing, it would be odd if it fixed one into a permanent state of ignorance. A second possibility would be to allow some kinds of information to overrule an estimate when they clash. But this evidently is not allowable. First, the kind of information with which a classical estimate conflicts is often more or less impressionistic. Thus the plausible opinion that the mean height of college students attending any London University college is unlikely to differ from that of the general population by more than twelve inches is not objectively founded on an analysis of random samples. Hence, it may not enter into a classical inference. But suppose a classical estimate is contradicted by objective information, say, from another classical estimate. It seems reasonable to combine the two estimates, giving greater weight to the one derived from the larger sample. But the classical method provides no mechanism by which different estimates may be combined to give a single overall estimate. Indeed, it precludes such combinations, for each separate estimate is supposed to be accepted as correct. The classical statistician therefore has no incentive to repeat a trial and would be quite unprepared to deal with the result of such a repetition.

Thus, classical estimation is prone to deliver estimates which prior knowledge would deem unreasonable and improbable, or even definitely false, and it seems to have no facility for revising estimates by taking account of such knowledge. The Bayesian approach, of course, meets no comparable difficulty, for it makes explicit provision for prior information.

iv The multiplicity of competing intervals. The standard account of confidence intervals requires one to accept unequivocally that a parameter lies in a specific region, this bold move being justified on the grounds that one will rarely accept a false theory. But besides the objections already levelled at this rule, which we regard as decisive, there is also the difficulty that intervals with a given confidence coefficient are not unique. Indeed, when the sampling distribution is continuous, there are infinitely many such intervals. This is easy to see from the diagram of a sampling distribution of means given earlier. Indefinitely many regions of that normal distribution cover 95 per cent of its area. For instance, one could consider an interval extending further into the tails of the distribution but omitting a small strip in the centre. The multiplicity of equally probable confidence intervals plays havoc with the type of conclusion recommended in classical statistics. Since every possible sample value falls into some acceptance region and also into some rejection region, unless the permissible confidence intervals were more strictly confined, we would have to "behave as if we actually knew" (*see* Neyman on page 190) that the parameter both has some value and fails to have any value!

Defenders of the theory of confidence intervals have tried to meet this seemingly disastrous objection in one of two ways, neither, we believe, satisfactory. The first looks to the length of a confidence interval as a way of choosing the one that is best. Intervals for a given confidence coefficient may vary in length, and it is frequently held that the shortest interval is best. Thus, in the earlier example of the estimation of a population mean, the centrally symmetrical interval given by

$\bar{x} \pm k\sigma_{\bar{x}}$ is always preferred since it is the shortest. However, it is hard to see how this preference could be rationalised, for the only justification offered for inferences based on confidence intervals, namely, that repeated inferences of that kind would rarely lead one into error, applies to broad and narrow intervals with equal force (though, as we have seen, 'force' is hardly the right word).

Cramér (1946, p. 513) held that it will "Obviously . . . be in our interest to find rules which, under given circumstances, yield as *short* confidence intervals as possible", and Hays (1969, p. 290) claimed that there is "Naturally . . . an advantage in pinning the population parameter within the narrowest possible range with a given probability". Neither spelled out exactly the supposed advantage of the narrowest confidence intervals, though they seem to be suggesting that by limiting inferences to them, one gains precision without sacrificing the associated confidence level. This was certainly Mood's view (1950, p. 222): in comparing two 95 per cent confidence intervals, he stated that one of them was "inferior" because of its greater length, for "it gives less precise information about the location of" the parameter. In a sense, this is correct. It is worth noting, however, that when the statistician accepts that θ lies in the range (a,b), he necessarily accepts that $f(\theta)$ lies in the range $[f(a),f(b)]$. And while the first interval might be the shortest in a set of intervals for θ, the second may not be the shortest in the corresponding set for $f(\theta)$. In other words, while the shortest confidence interval might deliver the most precise information regarding θ, the information it carries regarding some function of θ might be sub-optimal in precision. And since information about a parameter drawn from experiments is often used for both theoretical and practical purposes to calculate other quantities by means of some function, this is an objection of some weight.

Another difficulty is this: although the minimum-length criterion fixes a unique confidence interval for a given estimator, it leaves open which statistic should be employed for the estimation. The trouble is that different shortest confidence intervals may often be calculated from a variety of sample statistics. Clearly, the criterion based on the width of intervals needs to be supplemented by a further restriction, and it is usually suggested that the preferred statistic should have a minimum variance.

In justifying this new condition, it is argued that although confidence intervals constructed with a minimum-variance statistic are not necessarily the shortest in any particular case, such statistics have a higher probability than any other of producing the shortest interval. As many statisticians misleadingly express this conclusion, confidence intervals based on the preferred statistic "are shortest on average in large samples" (Kendall and Stuart, 1979, p. 126). We have already criticized both the long-run justification and the shortness criterion; since two wrongs don't make a right, nothing more needs to be said.

Neyman suggested a different criterion for selecting appropriate confidence intervals. He argued, in a manner familiar from his theory of testing, that a given confidence interval should not only have a high probability of containing the correct value, but should also be relatively unlikely to include wrong values. This is more precisely formulated as follows: a best confidence interval, I_o, should have the property that, for any other interval, I, corresponding to the same confidence coefficient $P(\theta' \in I_o \mid \theta) \leq P(\theta' \in I \mid \theta)$, where $\theta' \epsilon I$ means that θ' is a member of the set of points constituting the interval; moreover, the inequality must hold whatever the true value of the parameter, and for every other value θ' different from θ (Neyman, 1937, p. 282). But as Neyman himself showed, there are no so-called best intervals of this kind for most of the cases with which he was originally concerned. In particular, where the probability distribution of the experimental outcomes, governed by the unknown parameter, is continuous, no such best interval exists.

■ d CONCLUSION

The classical theory of estimation was designed to meet the need for more precise inferences from experimental data than can be got from tests of significance, and it was required to achieve this without reference to Bayes's Theorem and the inevitable element of subjectivity involved in assigning prior probabilities to hypotheses.

Point estimation supplies very precise estimates of parameters. However, those who advocate this type of estimation are normally rather vague about the type of conclusion that their methods are supposed to endorse. Although the statement that

t_o is the best estimate of θ seems quite definite, its exact meaning is hard to discern. Kendall and Stuart, who are among the few statisticians who have attended to this question, apparently held that a point estimate is the assertion that the true value and the estimated one are the same, for they claimed that a point estimate is a "number which can be taken to be the value of" the parameter in question. They made a similar claim for interval estimates. Indeed, if this were not the import of a classical estimate, it is difficult to imagine what would be. But nobody should, or, in our experience, ever does categorically assert, or definitely believe, or even act as if they believed, that which is almost certainly false. We have reviewed the various criteria which, if satisfied by a method of point estimation, are supposed to license such a conclusion. We have argued that they fail to do so and that insofar as two of the criteria (namely, sufficiency and efficiency) are plausible, this may be explained by reference to Bayes's Theorem.

Interval estimates are thought by most classical statisticians to get round the difficulty with point estimates that they lead to absolutely precise and, hence, almost certainly false estimates. But while an interval estimate is more realistic in giving a range of possible values for some parameter, the impression that it also provides a measure of the degree of confidence appropriate to the conclusion is illusory. Many classical statisticians appear to concede this when they say that an interval estimate must be unequivocally accepted as true, or at least that one should shape one's behavior as if one accepted it to be true. This interpretation, according to Kendall and Stuart, "is basic to the theory of confidence intervals" (1979, p. 110). But for the reasons we have given, we regard such advice as both unrealistic and unreasonable. It is particularly unreasonable because it may lead one to accept an estimate that other information suggests is very unlikely to be true.

In the light of these objections, we submit that the support enjoyed by classical methods of estimation among statisticians is unwarranted.

■ PART IV

The Bayesian Approach to Statistical Inference

In this part of the book we shall first consider the nature of the probabilities that statistical hypotheses purport to describe. This will involve a discussion of whether or not those probabilities represent, as some believe, states of the mind, or whether, as we shall contend, they are physical qualities of the external world. This will be the topic of Chapter 9. We shall then give an account of the manner in which statistical hypotheses may be evaluated by reference to their subjective probabilities, in accordance with Bayes's Theorem. We do not intend to provide a manual of Bayesian methods for dealing with every kind of hypothesis and prior structures of beliefs. A number of excellent works of this nature already exist. Our object, which we pursue in Chapter 10, is to show how, in general, the Bayesian approach works for inferences in statistics, and how it unifies the various intuitions that scientists have developed in regard to inductive reasoning—intuitions which are only accommodated, if at all, in the classical treatment by means of what we have argued are unconnected and unsatisfactory rules. In the final part of the book we shall respond to the numerous objections with which the Bayesian approach has been assailed.

CHAPTER 9

Objective Probability

■ a INTRODUCTION

Statistical inference is the experimental evaluation of statistical hypotheses, that is to say, hypotheses ascribing probabilities to types of event, such events usually being characterised by the values of one or more random variables, like the number of heads to appear in n tosses of a coin. We have already discussed the principal classical techniques for evaluating such hypotheses and have argued that they are thoroughly fallacious. Before we argue that the Bayesian approach succeeds where these fail, we must say something about the nature of the probabilities which statistical hypotheses talk about.

So far we have simply assumed that these probabilities characterise certain types of experimental situation without finding it necessary to give any finer analysis of what it means to say that a given experiment has a particular probabilistic structure. Because the probabilities purport to describe quite objective, structural features of experiments, we shall call them objective probabilities to distinguish them from the subjective, degree-of-belief probabilities introduced in Chapter 2. Some discussion of the nature of these objective probabilities is necessary, however, since merely saying that a particular experiment is characterised by an objective probability-distribution over its outcome space gives you no information concerning what, if any, empirical features you should look at in order to evaluate this claim.

We shall conduct, in this chapter, an examination of the extant accounts of objective probability. We shall show that all of them are vulnerable to powerful objections and that for all but one of them these objections appear to be quite decisive. We shall show that the sorts of criticism usually thought to be decisive also against this remaining theory (von Mises's) lose their force when a Bayesian framework for evaluating hypotheses about the values of those probabilities is adopted. That the Bayesian theory offers the only sound foundation for a logic of inductive inference was the burden of chapters 4 to 8; we shall reinforce this conclusion by arguing that only within such a framework do statistical hypotheses acquire any empirical significance at all.

The principal objection to the extant theories of objective probability is that they, or rather the statements within them which attribute particular values to events, do not seem to be either strictly verifiable or falsifiable. Some people, most notably de Finetti (1937 and 1970) and Savage (1956), conclude from this that their empirical content is nil and that all appeals to objective probabilities therefore merely import unwelcome metaphysics into empirical science. De Finetti and Savage have gone on to argue that statistics can function perfectly adequately, moreover, without making any reference at all to objective probability.

We shall discuss this last claim in due course. First let us look at the accounts of objective probability currently on offer. To provide a context to the discussion let us initially suppose that we are considering the hypothesis that relative to the 'experiment' of tossing some given coin, the outcome heads occurs with a definite probability p. We shall see in the subsequent sections how the varying accounts of objective probability propose to characterise the parameter p and what sorts of data they correspondingly regard as providing information about its true value.

■ b VON MISES'S FREQUENCY THEORY

This very well-known theory is of a type known as frequency theories of probability. What all such theories have in common is that they define probabilities relative to a sequence of outcomes of some repeatable set of conditions, like tossing the coin in the example above. Relative to such a sequence, the parameter p is regarded as being approximated by the ratio equal to the number of times a head occurs in the first n tosses of the coin, divided by n (this ratio is called the relative frequency of heads in the first n tosses).

Though the idea of characterising probabilities in terms of relative frequencies was announced in the nineteenth century (most notably by Venn (1866), frequency theories were developed in a detailed and systematic way only in this century:

Fisher (1922), Richard von Mises (1939, 1957, and 1964), and Reichenbach (1935) all set out such accounts, which differ from each other to a greater or lesser extent. The best known of these theories is undoubtedly von Mises's; and as it is also the one which we shall eventually adopt, we shall consider it in some detail. In particular, we shall note those of its characteristics which have drawn much, apparently unanswerable, criticism. We shall conclude, however, that while von Mises's theory cannot, as he presented it, furnish an adequate scientific theory of probability, it can when embedded suitably within the theory of subjective probability. Let us now see what are the principal features of von Mises's theory, and why they seem to pose such intractable difficulties.

b.1 The Collective

I The Axiom of Convergence. Von Mises's theory can probably best be presented as a relatively abstract mathematical theory coupled with an intended interpretation. The fundamental notion of the theory is that of a *collective*. A collective, considered (as in 1964, p. 12) in the first instance purely mathematically, is an infinite sequence $w = (w_1, w_2, \ldots, w_n, \ldots)$ of 'attributes' drawn from some set A, which satisfies certain characteristic conditions which we shall mention in due course. These attributes are intended to be a class of types of exclusive and exhaustive possible outcome—the outcome space, in other words—of what von Mises calls a "repeatable event", by which he means some state of affairs determined by a set of conditions which can in principle be repeatedly instantiated without limit. We shall use the term 'repeatable experiment' for 'repeatable event': there is a sense in which no event is repeatable, whereas we are accustomed to regarding an experiment as a prescription for bringing into existence states of affairs which are the same in some relevant class of respects. Throwing a die is one of von Mises's favourite illustrations of such a repeatable experiment; here, if it is just the number on the uppermost face which we take as our focus of interest, then A will be the set {1,2,3,4,5,6}. In the coin example we started off by discussing, the attributes would simply be the pair {heads, tails}. (The set A need not be finite, however; indeed, it may even be uncountably infinite. Random variables taking all real values are defined on an attribute space of the cardinality of the continuum, for example.)

The collective w is intended to represent a sequence of outcomes, characterised in terms only of which attribute from A they manifest, which would be obtained were a particular repeatable experiment, call it E, to be repeated indefinitely often. (There is, incidentally, a slight discrepancy between von Mises's own definitions of a collective; in the 1957 text of *Probability, Statistics and Truth,* a collective is described not as an infinite sequence of potential outcomes, but as a finite sequence of actual outcomes, which would satisfy the convergence axiom were it to be indefinitely extended. The difference seems only a nominal one, however.) The sequence of repetitions of that experiment or observation is not necessarily a temporally ordered sequence: the experiment consisting in tossing, in a specified way, a coin of specified dynamico-geometrical characteristics can be repeated either by n temporally successive tosses of one such coin or by simultaneous tosses of n coins satisfying the specification. Whether, if n could be allowed to tend to infinity, the resulting sequence in either case would form a collective will then depend on whether two conditions are satisfied.

The first condition is a condition on the relative frequencies with which each attribute in w is manifested. This condition is stated in von Mises's *Axiom of the Existence of Limits,* or *Axiom of Convergence,* as we shall call it for short. This says that the relative frequency of C among the first n members of w, for each n and each C in A, must tend to a definite limiting value as n tends to infinity. Thus, if E is the coin-tossing experiment, and A the set $\{H,T\}$, and if this experiment would generate a collective, were it to be indefinitely repeated, then the relative frequencies of heads and tails must, by the Axiom of Convergence, tend to definite limits in the collective.

Von Mises introduced the postulate of the existence of limiting relative-frequencies explicitly as the theoretical law corresponding to an apparently well-attested type of empirical regularity, which he termed the Empirical Law of Large Numbers. This 'law' expresses the fact, or the apparent fact, that on repeatedly performing those types of experiment which appear to generate their various possible outcomes 'at random', like tossing a coin or measuring the heights of people drawn 'at random' from a large population and recording the value to the nearest inch, the observed relative frequency of each type of outcome seems to converge within an increasingly small interval as the number of repetitions increases. (We put 'at random' in quotes because *randomness* has not yet been defined; a definition of it is part of von Mises's theory, however, as we shall see.) For example, a few hundred tosses of a particular coin are found to give a relative frequency of heads whose first decimal place does not change subsequently; some thousands more tosses determine the second decimal place, and so on.

This is von Mises's justification of his Axiom of Convergence. Observed relative frequencies in stochastic experiments appear to converge; and the propensity of those experiments to generate collectives with characteristic limiting relative frequencies is the fundamental theoretical law he proposed in explanation of this fact. It remains, of course, only a hypothesis that any given type of repeatable experiment has the property that it would generate such collectives in the counter-factual circumstances that it could be performed infinitely often. How one is actually to construct empirical tests of such a hypothesis is something we shall discuss in detail shortly. We still have not fully characterised the notion of a collective. The Axiom of Convergence tells only half the story; the remainder is expressed by von Mises's Axiom of Randomness. Let us now see what this says.

II The Axiom of Randomness. Consider again the example with which we commenced this chapter, that of the tossed coin. We have noted what appears to be an empirical fact, namely that the relative frequency of heads stabilises in the neighbourhood of some characteristic value. But we also observe, as we continue to toss the coin, that coexistent with the stability of the long-run relative frequencies of heads and tails, the order in which those two outcomes occur appears to defy all attempts at systematic prediction. And not only do those outcomes defy prediction in general, but there also seems to be no system of prediction which yields a higher or lower frequency of heads than the frequency of heads in the population, or sequence as a whole. Were such methods of prediction available, whose success rate differed from the frequency of the relevant character in the population, then they would form the basis for betting on or against at more favourable odds and would therefore constitute a gambling system.

To say that the outcomes of some stochastic device are

immune to gambling systems is therefore equivalent to saying that no system of prediction would, if indefinitely pursued, yield different frequencies of occurrence than those in the population, or collective, as a whole. This is the *Axiom of Randomness*: that the limits of the relative frequencies of all attributes are the same in any infinite subsequence of a collective w determined by any *place-selection,* or effectively executable method of selecting in advance indices i in w which will correspond to outcomes on which you will bet.

Whether i is selected by some given place-selection or not is allowed to depend only on i and a knowledge of the predecessors of w_i in w. A place-selection can therefore be represented as a function $f(s_n)$ of finite sequences of n attributes (representing possible initial segments of length n of w), and taking values 1 and 0, where 1 means 'select the $n + 1$th index of w', and 0 means 'do not'. That the place-selection is effectively executable is therefore the same as saying that f should be effectively computable. The highly developed theory of computable, or recursive, functions from natural numbers to natural numbers can be exploited, if the attribute space A is finite or denumerably infinite, by 'coding' any finite sequence of attributes by a unique natural number. A place-selection can now be defined to be any computable function $H(z)$ of one natural number argument z, and taking only the values 0 and 1, such that H selects w_{n+1} in w just in case $H(v_n) = 1$, where v_n codes the first n elements $(w_1, \ldots, w_n)$ of the collective w (an equivalent translation of the theory of place-selections into the theory of computable number-theoretic functions was first carried out by Church, 1940). To sum up: we can say that a collective w is random in the sense of von Mises–Church if there is no recursive place-selection which selects an infinite subsequence in which the relative frequencies of the attributes tend to limiting values which differ from those in w.

The von Mises–Church characterisation is not the only formal explication of some informal notion of randomness to have been suggested, nor is it the only one to have been based on the notion of a computable algorithm. Kolmogorov (1965), for example, employed a quite different approach, based on the observation that a computer program for generating any finite sequence of zeros and ones is longer, the more irregular the sequence. Unlike the von Mises–Church approach, this suggests the following criterion of randomness for *finite* sequences:

a sequence is random if the minimum length of computer program required to compute a sequence of given length exceeds some appropriate positive integer. The minimum length of program required to compute a sequence of specified length will depend not only on the sequence's irregularity, however, but also on the relative frequency of ones in it: clearly, as that relative frequency falls away on either side of $\frac{1}{2}$, the minimum length of program required to specify the sequence decreases. Because this is an account of randomness considered simply as intrinsic disorderliness, the minimum length of program above which the sequence computed by that program is classified as random is made to depend on the relative frequency of ones in, or the *weight* of that sequence.

Whether a sequence is random in this sense of exceeding a critical value of computational complexity, adjusted to relative frequency or not, will depend on the choice of that critical level and the particular type of computer chosen as standard. The complexity definition is therefore infected with arbitrariness from two distinct sources. A result of Kolmogorov's partially mitigates this arbitrariness, for it shows that for any pair of computers the complexity functions based on each differ by a constant. But the choice of critical level remains a matter of pure stipulation.

There are other attempted characterisations of randomness, for both finite and infinite sequences (Fine, 1973, and Earman, 1986, contain excellent discussions). They are strongly non-equivalent, however, in that no pair determines exactly the same class of random sequence; indeed, it seems highly doubtful that there is a anything like a unique notion of randomness there to be explicated. But there is nevertheless one notion of randomness which all statisticians employ when they talk about random samples, which is—or so it appears—defined purely probabilistically: a sample determined by the values of n random variables $X_1, \ldots, X_n$ is random if the X_i are independent and have the same probability distribution, that is, the probability of X_i's taking any value is the same as X_j's. The von Mises–Church definition of randomness, based on the idea of immunity to gambling systems, turns out to satisfy this condition: the successive outcomes of a collective are probabilistically independent with a common distribution. To this extent von Mises's Axiom of Randomness is entirely appropriate given the central objective of his theory, which was to provide a unified explanatory account of the scientific uses of probability. We shall discuss the connection between randomness and independence in von Mises's theory shortly; it is now time to give the explicit definition of probability with respect to a collective.

b.2 Probabilities in Collectives

If the two axioms, that of convergence of the relative frequencies of each type of outcome, and their invariance under all place-selections which select infinitely many members of w, are satisfied, then the limiting relative frequency of an outcome is called by von Mises its *probability* relative to the collective w. The use of the term 'probability' is formally justified, since von Mises's theory satisfies the axioms of the probability calculus as we presented them in Chapter 2. We shall now demonstrate this fact; those prepared to accept it on trust can skip to section **b.3.**

First, consider a type of experiment E with attribute space A. From statements describing these attributes we can, using the usual logical operations, generate a more extensive class of statements describing various more general types of outcome (which we may also describe in terms of the values of suitably defined random variables). For example, E might consist in measuring the position of a particle 'randomly' scattered onto a plane, with A obtained by partitioning the plane into small square intervals whose area is determined by the precision of measurement. The more extensive class of descriptions is then constructed by adjoining to the statements of the form 'the particle has attribute C', where C is the property of belonging to one of these elementary intervals, all other statements of the form 'the particle is in interval I', where I is any interval, and then closing off the extended class of statements under truth-functional operations. We shall show that if E would generate a collective if indefinitely repeated, then von Mises's probability function is defined on a subset of this extended set of sentences and satisfies the probability axioms.

Let us then assume E capable in principle of being repeatedly instantiated, and that were it to be so, it would generate a collective. Now consider the class H of statements h, describing types of outcome of E, such that the limiting relative frequency with which h would be satisfied in an indefinitely repeated sequence of instantiations of E exists and is unique (it is known, incidentally, that this class is not closed even under the ordinary truth-functional operations). Let the domain of P be H and let $P(h)$ be the limiting relative frequency with which any such statement h would be true.

It is very easy to see that each of the axioms 1 to 3 of Chapter 2 is satisfied by this interpretation, and we shall leave this to the reader. As for axiom 4, suppose that $P(h_2) > 0$. Let w be a collective generated by E, and w' the sequence composed of all the successive attributes instantiating h_2 in w. Since $P(h_2) > 0$, it follows that w' is infinite. It is fairly straightforward to show that w' is also a collective; that is to say, it satisfies both the condition of convergence and that of randomness. Define $P(h_1 \mid h_2)$ to be the limiting relative frequency with which h_1 is satisfied within the subsequence of those outcomes which satisfy h_2. It is easy to show that $P(h_1 \mid h_2)$ is well-defined if $P(h_1 \& h_2)$ is (i.e., the limits exist), and that it is then equal to $\dfrac{P(h_1 \& h_2)}{P(h_2)}$. For let $n(h_1 \& h_2)$ be the number of instances of $h_1 \& h_2$ in the first n members of w, and similarly for $n(h_2)$. Let $n'(h_1)$ be the number of instances of h_1 in the first n' members of w'. Then

$$\lim \frac{n(h_1 \& h_2)}{n(h_2)} = \lim \frac{n'(h_1)}{n'} .$$

But

$$\frac{n(h_1 \& h_2)}{n(h_2)} = \frac{n(h_1 \& h_2)}{n} \div \frac{n(h_2)}{n}$$

and by assumption the limits of both terms on the right-hand side exist and the second is positive. Hence

$$P(h_1 \mid h_2) = \frac{P(h_1 \& h_2)}{P(h_2)} .$$

We can note that the principle of countable additivity is not true in general, however. A simple counter-example is provided by Giere (1976, p. 326), as follows. Let the attribute space be denumerably infinite, of the form $\{A_1, \dots, A_n, \dots\}$. Suppose that there is a collective of these attributes and that the ith member is A_i. Then the limiting relative frequency of A_i, for each i, is the limit of the sequence $\dfrac{1}{n}$, as n tends to infinity, which is 0, and the limiting relative frequency with which the disjunction 'A_1 occurs or A_2 occurs or ...' is satisfied is obviously 1. But each of the disjuncts has probability 0 and a denumerable sum of zeros is 0.

Van Fraassen regards facts such as this, and the nonclosure of the domain of P under countable and even finite disjunction and conjunction, as providing strong reasons against characterising probability—at any rate as it is used in science—as a limiting relative frequency (1980, pp. 184–185). His argument is that science makes extensive use of probabilities whose domains are not only fields but also sigma fields (see Chapter 2 section **c** for an explanation of the notion of a sigma field). Moreover, those probabilities are countably additive, because they are often 'geometric', that is to say defined for events of the form 'the outcome is in Q', where Q is a measurable region of n-dimensional Euclidean space, and made proportional to the 'volume', or Lebesgue measure of Q (Lebesgue measure is an extension of the ordinary length, area and volume measures for intervals in Euclidean space of one, two, three, etc. dimensional space to a wider class of sets, known as the Borel sets, generated from these intervals). Since Lebesgue measure is countably additive, any probability measure P proportional to it is also countably additive; for if Q above is the union of a countable family of pairwise disjoint sets Q_i, then $P(Q) = k\lambda(Q) = k\lambda(\cup Q_i) = k\Sigma\lambda(Q_i) = \Sigma k\lambda(Q_i) = \Sigma P(Q_i)$, where $\lambda(Q_j)$ is the Lebesgue measure of Q_j. So where probability is proportional to Lebesgue measure, then it is countably additive.

But this argument does not show that in every type of scientific application probability must be countably additive, only that that it is in those contexts where proportionality to Lebesgue measure is deemed necessary. Also, there is the consideration that smooth mathematical theories, like that of countably additive probabilities defined on sigma fields, might owe their acceptance into science precisely to their being smooth, well-understood mathematical theories. Indeed, van Fraassen's arguments rest on taking such theories at their face value, rather than regarding them as justified on largely pragmatic grounds. We see this again in his next argument against a limiting relative frequency account of probability, in which he shows (1980, p. 185) that limiting relative frequency probability spaces can never be identified with 'geometrical' probability spaces (a probability space is essentially just the probability function and its associated attribute space). He considers the example of a dart's being thrown 'randomly' at a dart-board. Idealising, suppose it hits at a mathematical point each time, and that the throws generate a collective. Let A be the set of points hit in denumerably infinitely many throws. Hence the limiting relative frequency of the event 'the hit is in A' is 1. Given the 'randomness' of the throws, it is plausibly assumed that the appropriate probability model is one in which the probability of a hit in any region of the board is proportional to the area of the region. But then the probability of a hit in A is zero, since the Lebesgue measure of a denumerable set is 0. Also, the limiting relative frequency of a hit in the complement of A is zero, while its 'geometrical' probability is clearly 1.

For a conventionalist (as is van Fraassen), examples like this represent a decisive objection to a limiting relative frequency theory of probability. But a realist should no more feel forced to make a choice between limiting relative frequency and 'geometrical' probabilities than he should between a belief that a piece of matter is discrete and the adoption of a continuous function to describe its mass distribution. Continuous mass distributions are a fiction, but a useful one, for they offer a mathematically simple approximation to the true state of affairs. Likewise, 'geometrical' probability distributions are a sufficiently good approximation to appropriate relative frequency distributions to warrant their use as simple mathematical models of them. Consider, for example, a standard 'geometrical' probability distribution, in the kinetic theory of gases. For those subsets of the phase space of a gas which physicists want to consider, the relative frequency, in a collective of gas samples in identical macroscopic states, of phase points belonging to those sets is thought to be closely approximated by their measure—or at any rate, it is thought so by those who take the view that the probabilities involved are objective. It is only fair to add that many people do not take that view; who is right in this, it is not for us to say. What we *can* conclude, however, is that van Fraassen's objections to relative frequency definitions of probability will deter only those who mistake the mathematical appearances of science for its substance.

b.3 Independence in Derived Collectives

We said earlier that an important consequence of von Mises's theory is that the successive manifestations of any n types of outcome in the collective are independent with constant probability. It of course follows that (among others) the Weak Law of Large Numbers is satisfied. This might seem rather surprising: the collective w represents successive outcomes of an indefinitely repeated experiment E, whereas the weak law refers to the successive outcomes of the derived experiment $E_{(n)}$ consisting of repeating E n times. A sequence of successive outcomes of the n-fold experiment $E_{(n)}$ is, of course, a sequence each member of which is an n-fold sequence of outcomes of E. But w is a sequence of single outcomes of E, and therein—or so it seems—lies the problem.

The solution to the problem lies in noting that any sequence can be partitioned into a sequence of pairs, of triples, etc.; the difference is at the descriptive level only (the logician would say the difference is one of type-level). For example, we can partition the natural number sequence into segments of length 4:

$$((0,1,2,3), (4,5,6,7), (8,9,10,11), \ldots).$$

By partitioning w into successive segments of length n, therefore, that very same sequence can be regarded as representing an infinite sequence of outcomes of $E_{(n)}$. It is not very difficult to show (and it is shown in von Mises, 1964, pp. 27–28) that if w is a collective, then so is $w_{(n)}$, the sequence of n-tuples of attributes in A obtained by partitioning w. Now suppose that B is some property of outcomes of E. If we define n binomial random variables X_i on the members of $A_{(n)}$, the set of all n-tuples of members of A (and written A^n in set-theory textbooks), such that X_i takes the value 1 on $(C_1, \ldots, C_n)$, if C_i is an instance of B, then it can be shown that the convergence and randomness properties of w together imply that the X_i are independent with the same probability, relative to $w_{(n)}$, of being equal to 1 as the probability of B relative to w (the proof of this result is very straightforward and is found in von Mises 1964, pp. 28–31). Thus a collective w with attribute space $\{0,1\}$ can be partitioned sequentially into a collective of Bernoulli trials with probability parameter p equal to the probability within w.

At first sight this might appear to prove too much. After all, a great deal of applied probability theory deals with sequences of events which are not independent, like those forming Markov chains, for example. This type of application seems to be precluded if the account we have just given is correct. It is not difficult to see, however, that this conclusion is not correct. For consider $E_{(n)}$, where E is the experiment of tossing a coin once. $A_{(n)}$, the attribute space of $E_{(n)}$, is the set of the 2^n n-fold sequences of H's and T's. Suppose E would generate a collective w were it to be indefinitely repeated. Divide w successively into n-fold sequences to give $w_{(n)}$. Let $X_1, \ldots, X_n$ be n random variables defined on the attributes in $A_{(n)}$ such that $X_i = 1$ on a given sequence if the ith member of that sequence is a head. Then the X_i are independent (given that E does in fact generate a collective). Let Y_m, $0 < m \leq n$, be the n random variables which record the number of heads in the first m members in those sequences. Then the Y_m are certainly not independent, though the X_i are. In other words, it is quite possible to construct dependent random variables within this theory in which successive attributes in the collective are independent.

b.4 Summary of the Main Features of Von Mises's Theory

Before we proceed to the next main topic of discussion, whether von Mises's theory does, as he claimed, furnish the foundation for scientific applications of probability, we shall summarise the main features of that theory.

1. Some—and it is the task of science to discover which—repeatable experiments E, with associated attribute spaces A, have the property that, were they to be repeated indefinitely often, they would generate collectives of members of A.
2. Within these collectives the relative frequencies with which the attributes occur tend to limits (Axiom of Convergence), and limits which are invariant under place-selections (Axiom of Randomness).
3. A class of subsets of A, which includes all the singleton members of A, corresponds to a set of more general descriptions of possible outcomes of E. For any given collective of members of A, let H be the class of those descriptions h such that $P(h) = \lim\dfrac{n(h)}{n}$ exists in w. Then P satisfies the probability calculus, where the domain of P is H.
4. H is not closed under truth-functional operations. Moreover, though $P(h) = \lim\dfrac{n(h)}{n}$ is finitely additive, it is not countably additive.
5. Within the derived collectives $w_{(n)}$, the successive outcomes of E are independent with the same probability of occurrence as in w.

Let us now leave the exposition of von Mises's theory and assess its credentials of scientific adequacy.

b.5 Is Von Mises's Theory a Good Theory?

We have already observed that according to von Mises many

types of repeatable experiment generate collectives, or at any rate would do so if they could be continued indefinitely. The task of statistics, for him, is to identify which experiments have this collective-generating property and to elicit the associated probability distributions over their class of possible outcomes. Two observations are appropriate at this point.

We noted earlier the rather obvious fact that no collective exists or will exist in any real sense. It is a postulated ideal object, conceived only via the experiment which by assumption generates initial segments of it in any actual sequence of repetitions. It does not follow, however, as some have charged, that von Mises's theory is necessarily either meaningless or unscientific because of the strongly counter-factual way in which its basic entities are characterised. Classical mechanics is paradigmatically scientific, yet it tells you what would happen if, for example, two perfectly elastic spheres, travelling with given velocities, were to collide. There are no perfectly elastic spheres; nor are there ideal gases, nor are there point masses, nor are there many other of the entities whose existence appears to be postulated by science and which furnish the basic entities of its theories. Science accommodates them because their postulation leads to scientifically successful theories.

The question which we shall now try to answer is whether von Mises's account justifies the considerable idealisation involved in its basic assumptions and yields a successful scientific theory of the domain of stochastic phenomena. Unfortunately, scientific methodology—and in particular, the issue of what is the essence of a good theory—seems to be an area where there is much in the way of individual taste expressed and relatively little in the way of generally accepted criteria against which to evaluate these tastes. It is very often claimed that good theories are explanatory of the phenomena in their domain; but anybody who has ever investigated the extraordinarily extensive literature on the topic of explanation in science must have been struck by the inconclusive nature of a discussion which has been conducted by philosophers over many decades, not to say centuries. We wish the discussion to be as little dependent as possible on the vicissitudes of the discussion as to what truly constitutes an explanation; and we shall accordingly rest content with judging von Mises's theory against one minimal standard on which practically everybody agrees (though there are exceptions even to this): empirical testability.

Are hypotheses about the existence and magnitudes of limiting relative frequencies in collectives empirically testable?

b.6 The Empirical Adequacy of Von Mises's Theory

Unfortunately, the answer most people have arrived at, and for what appear to be very good reasons, as we shall see, is 'no': the theory seems to be pure metaphysics. On a limiting relative-frequency interpretation of probability statements, a hypothesis of the form $P(h) = p$ makes no empirically verifiable or falsifiable prediction at all, for it is well known that a statement about the limit of a sequence of trials hypothetically continued to infinity contains by itself absolutely no information about any initial segment of that sequence. In other words, any initial segment of a collective—and we are, of course, only ever capable of observing initial segments—can be replaced with any arbitrary sequence of the same length without affecting any of the limits in the collective. Von Mises was perfectly aware of this apparently rather devastating objection to his theory, however, and attempted to evade its force by means of a variety of arguments. Let us consider these in turn and attempt to judge to what extent he was successful.

I The fast-convergence argument. This is the weakest of von Mises's responses to the charge of empirical emptiness against his theory. It is that one in practice makes the "silent assumption" that the convergence to the limit in a collective is fairly quick (von Mises, 1964, p. 108). The obvious defect of this reply is that it is hopelessly vague and does not allow us to decide whether obtaining, for example, 273 heads out of 500 tosses of a coin provides us with any definite information as to whether the coin is fair. Clearly, without some method of making such decisions, the theory of collectives does seem to remain a piece of untestable metaphysics.

II The laws of large numbers argument. Von Mises invokes limit theorems of mathematical statistics, like Bernoulli's, about indefinitely long sequences of independent trials, as the means of making the "silent assumption" a little more audible and definite. He asserts that these limit theorems provide information which, in relating properties of initial segments of collectives to the limits of the relative frequencies within them, provide the required information about the speed with which convergence to the limits occur. Thus he claimed that "Poisson's Theorem [a variant of Bernoulli's in which the assumption of constant probability is relaxed] has a practical meaning for real sequences of observations" (1939, pp. 170–71), and he described Bernoulli's Theorem as "a very important theorem of practical statistics" (1939, p. 189).

Von Mises regarded such laws-of-large-numbers results as giving information about the rate at which convergence proceeds within a collective, for reasons which seem to be implicit in the results mentioned in section **b.3**. For suppose that an experiment E generates a collective w with attributes 0 and 1, and that X is a random variable taking the value 1 on 1 and 0 on 0, such that $P(X = 1) = p$. Then $w_{(n)}$, the sequence obtained by partitioning w sequentially into sequences of length n, is also a collective; and, moreover, the random variables $X_1, \ldots, X_n$ defined in section **b.3** on these sequences are independent and such that for each i, $i = 1, \ldots, n$, $P_{(n)}$ $(X_i = 1) = p$, where $P_{(n)}$ is the probability relative to $w_{(n)}$. Where $Y = X_1 + \ldots + X_n$, it follows that Y has the binomial distribution with probability parameter p. Hence the probability that Y lies within, say, 3 standard deviations of np is, for large n, close to 0.99, where the 3 standard-deviation points are $np - 3\sqrt{np(1 - p)}$ and $np + 3\sqrt{np(1 - p)}$. If E is tossing a coin once, and this generates a collective w in which $P(H) = \frac{1}{2}$, then in the derived collective of sequences of 100 tosses of the coin, the probability that the relative frequency of heads lies between 35 and 65 is close to 0.99.

But such a result unfortunately gives no information whatever about how fast convergence actually proceeds within w. For the meaning, within von Mises's theory, of the statement

$$P_{(n)} \ (Y \text{ is within 3 s.d.'s of } np) \approx 0.99,$$

where 's.d.' stands for standard deviation, is merely that in the collective $w_{(n)}$, the relative frequency of attributes (n-fold sequences of zeros and ones) in which Y is within 3 standard deviations of np tends to some number in a small neighbourhood of 0.99. That statement does *not* assert that if you perform E, n times in succession, you will with overwhelmingly large probability get a value of Y within 3 standard deviations of np; indeed, such an interpretation of that inequality is, as von Mises himself stressed in another context (that of so-called single-case probabilities, which we discuss in section **c**), totally illegitimate: $P_{(n)}$ refers, and refers only, to a limiting relative frequency of attributes within $w_{(n)}$.

It might be objected against this deflating conclusion, that you may confidently expect a value of Y lying outside the 3 standard deviation limits not to occur in any sequence of n repetitions of E, if the hypothesis, call it H, that E generates a collective in which 1 has the probability p, is true. For such an outcome would, by virtue of its small limiting relative frequency in $w_{(n)}$, be extremely unlikely to occur this time: there are, if H is true, so very few of them that for this outcome to be among them, an enormous coincidence would have to occur. If we were to observe such an outcome would we not, therefore, be justified in regarding it as prima facie evidence against H?

The answer is 'no'. First, we have to remember that von Mises's theory entails that if an outcome C of an experiment E has any finite probability, *however small*, then in an infinite sequence of repetitions of E, C must occur *infinitely* many times. To say that there are only a "few" occurrences of an outcome with a small probability has, therefore, a distinctly Pickwickian ring. Secondly, we have only to choose n large enough and we will *always* obtain a type of outcome of $E_{(n)}$ which occurs with arbitrarily small relative frequency in $w_{(n)}$: for in $w_{(n)}$ the limiting relative frequency of any given value of Y tends very quickly to 0 as n tends to infinity (assuming that $0 < p < 1$). "Enormous coincidences" therefore occur literally everywhere in $w_{(n)}$ for sufficiently large n, which means that without further information, no significance can be attached to the occurrence of any one of them. Indeed, phrases like "so very few" and "enormous coincidence" are no more than rhetoric and ought not to conceal from us that all one is justified in inferring from an occurrence of C to which H assigns a small probability is that an event has occurred to which H assigns a small probability.

iii The limits - occur - elsewhere - in - science argument. Von Mises pointed out, in a further attempt (1939, p. 124) to rebut objections to the apparent empirical emptiness of his limit definition of probability, that the use of limits is ubiquitous in science. Velocity, acceleration, density, work, for example, are all defined in terms of limits of infinite sequences, since the first three are derivatives and the fourth is an integral; and integration and differentiation are defined either as limits (or in terms of limits if there are complex quantities involved) of sequences of real numbers. Furthermore, nobody complains that the introduction of these notions renders those parts of science in which they appear metaphysical. The problem of how to relate what you observe in 500 trials of a putatively collective-generating experiment to the postulated frequency limits "is exactly similar to that occurring in all practical applications of theoretical sciences" (ibid.).

But this is not true. The difference, from the point of view of empirical significance, between, say, classical kinematics and dynamics based on velocity and acceleration and von Mises's probability theory is exactly that the former straight-forwardly generate testable hypotheses while the latter does not. The Galilean law of free fall, which asserts that the ac-celeration of a freely falling body is constant, predicts distance fallen after time elapsed, and both these quantities are straight-forward observables. Statistical hypotheses incorporating von Mises's probabilities do not, we should be aware by now, de-ductively predict *any* observable state of affairs at all.

But perhaps von Mises can make valid his analogy with the physical sciences, where after all few if any of the principal theories make any prediction directly about observables. These theories are saved from relegation to the status of metaphysics because they are enabled to make such predictions through being coupled with suitable auxiliary hypotheses. Why not here also? Gillies (1973) has proposed that probabilistic hypotheses are rendered empirically significant in just this way. Although Gillies does not define objective probabilities as frequency lim-its in collectives (he claims that probabilities should simply be regarded as primitive), he suggested a type of auxiliary hy-pothesis which might serve also in von Mises's theory to render its hypotheses testable.

Gillies's rule.   Gillies's proposed auxiliary hypothesis says that that part of the range of a random variable X which has a sufficiently small probability density, according to some hy-pothesis $H(X)$, is effectively prohibited to X, if $H(X)$ is true, the prohibition being subject to certain conditions being satisfied by the form of the density function (1973, pp. 171–72). Gillies calls this stipulation a Falsifying Rule for Probability State-ments; by invoking it, he claims to have made probabilistic hypotheses effectively amenable to the empirical check of possible falsification. His rule, being a rule, might not seem like a factual statement at all and therefore not like an auxiliary hypothesis as normally understood. Gillies claims nevertheless that it is analogous to statements like 'the mass of the Earth can be ignored in comparison with the Sun's in certain calculations of gravitational attractions'. Be that as it may, the falsifying rule Gillies proposes leads, unless qualified in some way, to inconsistencies (as Redhead, 1974, demonstrates) and also represents just that theory of significance testing which we have already examined and rejected as a legitimate way of evaluating statistical hypotheses. We therefore conclude that as it stands, Gillies's proposed rule is unacceptable.

The Kolmogorov-Cramér rule. Kolmogorov proposed, in his classic 1933 work, a weaker, less categorical type of auxiliary assumption. While he did not attempt to define the probability function directly, he stipulated that it is "practically certain" that if the repeatable experiment, characterised by a probability-distribution P over its class of outcomes, is repeated "a large number of times", then the relative frequency of any event A "will differ very slightly from $P(A)$". Cramér (1946) repeats this account using almost exactly the same words and refers to it as the frequency theory of probability.

Without further information about what the phrases in quotation marks are supposed to mean, however, this stipulation is far too vague to be of any use. 'Almost certain' possesses a quite precise meaning in mathematical probability theory. It means 'with probability 1', and Kolmogorov's and Cramér's theory sounds rather like a paraphrase of the Strong Law of Large Numbers, that with probability 1 the relative frequency of outcomes A in a sequence of independent trials with constant probability p tends to p. But for this to be informative would presuppose that we already know what 'with probability 1' means.

Kolmogorov and Cramér do provide a further empirical criterion, which is, incidentally, independent of the original one. This is that in a single performance of the experiment, we can be practically certain that any event with a probability close to zero will not occur. What they seem to mean by this statement is that you are entitled to regard a hypothesis as refuted by sample data if it ascribes a very small probability to an event instantiated by that sample.

This, however, is just Cournot's rule again. We pointed out in our earlier discussion of this (Chapter 5, section **a**) that it is very obviously unsound: by tossing a coin enough times one can generate an event—a particular sequence of heads and tails—which has an arbitrarily small probability whatever the true probability may be of heads; according to the rule, then, every hypothesis about the value of that probability is rejected, which contradicts the assumption that one of those values is the true one.

b.7 Preliminary Conclusion

Von Mises does not seem to have been able to rebut at all successfully the charge that his theory fails the minimal condition for being regarded as an adequate scientific account of random phenomena, namely that it permit an empirical evaluation of its constituent hypotheses. However, his theory does have one very important consequence which seems to describe a central feature of random phenomena and which any successful account, therefore, will also have to accommodate. The paradox is that its ability to generate this consequence seems also to be the very undoing of von Mises's theory, for the consequence in question is that random phenomena defy any attempt at categorical prediction. Had von Mises's theory, for example, predicted that within some specifiable number of trials the relative frequency of a particular type of outcome would be confined within some proper subinterval of the unit interval, then that would have been a ground actually for rejecting the theory; for it seems to be the case that once in so many hundred, or so many thousand, or so many million times, these bounds are exceeded, a fact (and it is a fact—statistical physics has the capacity to 'observe' in the behaviour of gases such enormous experiments) which the theory of sequences of independent random variables seems in some sense to explain since it assigns a small but finite probability to such rare deviations. The trouble with any suggestion that departures from large-scale regularity ought to be explicitly accommodated by any adequate theory, however, is again that the suggestion is accompanied by no systematic account of how these probabilistic 'predictions' are to be empirically evaluated. We shall, we hope, be able to provide such an account; before we do this let us first look briefly at the other theories of objective probability and see whether they succeed where von Mises (apparently) fails.

■ c POPPER'S PROPENSITY THEORY, AND SINGLE-CASE PROBABILITIES

c.1 Popper's Propensity Theory

In a series of papers between 1950 and 1970, Popper introduced what he called the "Propensity Interpretation" of the probability calculus. According to one characteristic statement of this interpretation (1959), certain types of repeatable experiment are endowed with dispositions or *propensities* to produce fixed limiting relative frequencies of their various outcomes, were they to be continued indefinitely under similar conditions. Popper claims that his thesis, that these propensities "are not only as objective as the experimental arrangements but also *physically* real" (1957, p. 69; his italics), arises from the consideration of an allegedly powerful objection, which we shall discuss shortly, to what he calls "the frequency theory", under which heading it is clear that he includes von Mises's theory. Before we discuss this objection, first compare this quotation from von Mises with that of Popper's

> the probability of a 'double six' [at dice] is a characteristic property of a given pair of dice (or it may be a property of the whole method of throwing) and is comparable with all its other physical properties. *The theory of probability is only concerned with relationships existing between physical quantities of this kind.* (von Mises, 1939, p. 18; our italics)

That probabilities are physical properties of experiments, manifested in a disposition to generate constant limiting frequencies, was for von Mises no less than "the 'primary phenomenon' *(Urphänomen)* of the theory of probability" (ibid). Indeed, how could probabilities, characterised as relative frequencies which "would tend to a fixed limit if the observations were indefinitely continued" (von Mises, 1957, p. 15), be anything other than dispositional properties of observations of some specified type—in other words, of an experiment of specific type? The fact is that this theory of objective probabilities, as dispositions of repeatable experiments to generate convergent relative frequencies when repeated indefinitely, is von Mises's own.

Popper nevertheless did proceed beyond von Mises in one respect. In his *Logic of Scientific Discovery* (1959a), Popper argues that since probabilities, characterised as dispositional properties of some experiment, appear to depend only on the experiment, these same probability-values may be attributed to predictions about the outcomes of *particular* occurrences of the performance of the experiment:

> This [dependence of the probabilities on the experimental conditions] allows us to interpret the probability of a *singular* event as a property of the singular event itself, to be measured by a conjectured *potential or virtual* statistical frequency rather than by the *actual* one. (1959a, p. 37; his italics)

These putative probabilities are therefore to be considered as *objective single-case probabilities*.

Von Mises not only did not make this step of transferring the collective's probabilities to the single case, but explicitly denied that it made any sense to attribute probabilities to single events at all. We shall see that von Mises had very good reasons for his refusal to extend the theory to the single case, but now we must consider the objection of Popper's, to which we referred above, and which has often been regarded as demonstrating a fundamental weakness in a frequency theory of probability and vindicating Popper's own propensity theory. We shall see that these conclusions are unfounded.

c.2 Jacta Alea Est

Popper's argument exploits a simple example. Two throws of a fair die are interpolated into a long sequence of throws of a heavily loaded die. Popper claims (1959a, pp. 31–35) that we should want to say that the probability of a six occurring on each of the two occasions of a throw of the fair die is not equal to the probability of a six as estimated by the long-run relative frequency in the sequence in which those throws actually occur, and he goes on to urge this as a criticism of "the frequentist", who "is forced to introduce a modification to his theory". The modification is to make the probabilities properties of the structural features of the experiment rather than of a particular sequence of events (1959a, p. 35).

However, if "the frequentist" to whom Popper addresses these remarks is von Mises, then the latter is certainly not forced to modify his theory because of Popper's observation. Von Mises would, as we have seen, have agreed with Popper that the probability of a six for the fair die is not the same as the probability of a six for the loaded die, and like Popper's, his justification of this claim would be that were the fair die to be tossed indefinitely often, the relative frequency of heads would tend to a limit equal to one sixth. But that would be the extent of his agreement, however; he would not have agreed that these observations justify, as Popper himself goes on to allege, the attribution of the probability one sixth—or any other value—to the outcome of either of the two throws of the fair die. As we observed, von Mises explicitly denied that such attributions were possible, and his denial is well-founded.

There is an age-old, and quite insuperable, objection to Popper's attempt to define single-case probabilities in this way. The objection rests on the observation that the outcome of a particular throw of the die, to return to Popper's example, is influenced in practice by a number of factors, *only one of which is the mass distribution of the die itself.* Variations in air density, convection currents, the strength and point of contact of the initial impulsive force, the distance above the point of landing at which the die is cast, and so on, also play a part in determining on which face the die will fall. Suppose that the face on which the die falls is uniquely determined by parameters $q_1, \ldots, q_k$. Suppose also that at the first throw of the fair die, these take values $q_{1_0}, \ldots, q_{k_0}$, and that the outcome of that throw is a four. Consider the experiment E_0 consisting in throwing the die in such a way that the q_i take the values q_{i_0} (no matter that E_0 may be impossible to perform in practice, because of the limits to which precise measurements can be made; it is an issue of principle, not practice, which is in question here). Clearly, the relative frequency of a six in any sequence of repetitions of E_0 is 0, whereas the relative frequency of a six in a long sequence of repetitions of the experiment E, specified merely by the condition that the fair die is thrown in the usual way, is one sixth. Now the first throw of the fair die in Popper's imagined hybrid sequence is an instance both of E_0 and of E. Hence the single case probability of a six at the first throw of the die is both 0 and one sixth, and we have a contradiction.

This sort of objection to Popper's attempt to apply statistical probabilities to single trials is so standard and so well-known—it is the old problem of the reference class, or population, or experimental conditions not being uniquely determined by the outcome space—that it is almost incredible that he should not have anticipated it. He was misled into his theory of single-case probabilities by his inference that since probabilities are properties of experiments, and the experiment is the same (trivially) at every repetition of it, then the probability must be the same at each repetition. As we see, however, this inference is a non sequitur.

Popper believes that his propensity theory offers a solution to some of the interpretative problems of quantum theory. There has been a keen interest expressed in Popper's theory for this reason. Objections based on purely physical considerations have also been brought against it, however, and Milne (1986) has also provided a very elegant and concise demonstration that Popper's theory cannot, in particular, provide a coherent explanation of the interference of probability waves in two-slit experiments, as Popper had claimed it could.

But Popper's idea, if not the details of his theory, of single-case propensities nevertheless appealed and continues to appeal to many people. It was subsequently taken up by Giere, who seems to have been the first to point out (1973 p. 473) that Popper's theory founders on the objection that any actual observation of some effect may instantiate many different probability-determining experiments, and whose own response to that objection is to restrict a Realist interpretation of single-case probabilities to those examples, which can be tentatively identified only in the quantum domain, of irreducibly indeterministic phenomena. Giere reserves a purely instrumental, as-if interpretation, of the theory for the macro-world, where in practice the 'propensities' for particular trials to yield their possible outcomes are obtained from the most specific reference class for which statistics exist; "but it must be realised that this is only a convenient way of talking and that the implied physical probabilities really do not exist" (1973, p. 481).

Giere diverges sharply from Popper in denying any necessary link between propensities and long-run relative frequencies. For Giere propensities are simply measures of "causal tendencies [in single trials] not reducible to relative frequencies whether actual or possible" (1976, p. 327). There are two objections to this account of single-case probabilities. First, the authentic domain of applicability of the theory—irreducibly indeterministic trials—is uncertain (the micro-world might turn out to be deterministic, however unlikely an eventuality that would seem today), and very restricted relative to the standard applications of statistical theory. Our objective is to find a theory of objective probability that will fit in with the practice of statistics, not to truncate the latter because we have found a theory which does not.

The second objection is more fundamental and seems to be unanswerable. Von Mises's theory may seem stubbornly deficient in empirical content, but the present account is, if anything, even worse. For Giere's single-case propensity theory conveys no information of any sort about observable phenomena, not even, we are told, about relative frequencies in the limit. It is not even demonstrable, without the frequency link, that these single-case probabilities are probabilities in the sense of the probability calculus, so their place in a discussion of objective *probabilities* is, in default of further development, based on nothing more than assertion. Nor do the limit theorems of the probability calculus help. How, without further information, is one to give empirical meaning to phrases like 'with probability one', or 'independent, identically distributed random variables'?

Others besides Giere and Popper have attempted to construct a satisfactory theory of single-case propensities. There are too many variants of the Popper-Giere approach to consider in detail, and we believe that none survives the main criticisms that we have brought against Popper's and Giere's own theories. But there is a rather different type of single-case theory, the theory of objective chance, so-called, proposed by Mellor (1971), Levi (1980), Lewis (1981), and Skyrms (1984), which purports to be both empirically significant and internally coherent. Let us now see what this theory says, and whether its claims can be sustained.

■ d THE THEORY OF OBJECTIVE CHANCE

This theory is most simply presented in the familiar context of the tossed coin. The coin is credited by the advocates of this theory with a characteristic chance of yielding heads when tossed, the chance being present whenever the coin is tossed in the prescribed way. The chances of successive outcomes are, moreover, independent. The chance distribution, with its characteristic independence property, is, in other words, thought of as a permanent characteristic of the coin, a piece of dispositional behaviour as much a function of its physical structure as its resistance to bending forces.

The adverse criticisms levelled against Giere's similarly non-frequentist theory of objective single-case probabilities seem to have been met by the advocates of this theory. In particular, they have an answer to the question why chances should obey the probability calculus. This (rather ingenious) answer is that chances license numerically equivalent fair betting quotients, and these, by the same sort of reasoning based on the Dutch Book Argument which we gave in Chapter 3, satisfy the probability calculus. Hence so must chances.

We shall discuss the adequacy of this answer shortly. Let us go on to the question of why the chance of heads from toss to toss is plausibly regarded as constant. According to Mellor, it is a propensity of the coin, when it is tossed, to give a particular chance of landing heads, that chance depending on the mass distribution of the coin; and this remains (very largely) the same from toss to toss. The independence of the successive tosses appears also to be a highly plausible assumption. A head at the ith toss does not, we seem justified in asserting, causally affect the possibility of a head at the $(i + 1)$th.

It would appear to follow from these considerations that the subjective probability of r heads out of n tosses of the coin, given that the chance of heads is p, is equal to $p^r(1 - p)^{n-r}$. So we can, it seems, simply insert this as the value of the likelihood term $P(e \mid h)$ in Bayes's Theorem, where P is our subjective probability function, e is the sample data, and h the given chance distribution, to compute, in conjunction with our prior degrees of belief $P(h)$ and $P(e)$, a posterior degree of belief in h. In other words, we appear to have a simple method of evaluating hypotheses about the values of chances from the data.

The principle, or axiom, that chances license numerically equivalent degrees of belief, called the *Principal Principle* by David Lewis (1981), is thus a very powerful one, for it implies that chances are probabilities in the sense of the probability calculus, and as we saw in the previous paragraph, that same principle enables us also, together with some plausible background information about the causal structure of the experiment, to evaluate the likelihood terms in Bayes's Theorem and thereby to evaluate empirically hypotheses about chance—for example, the hypothesis that the chance is actually constant from toss to toss. The theory, in other words, seems to supply, mainly in the Principal Principle, the deficiencies of von Mises's as well as of the single-case theorists' accounts.

But this is not really the case. For the Principal Principle is a synthetic statement relating chance to degree of belief which , however, receives no independent justification. According to Lewis, the Principal Principle contains *all* that we know about the structure of chances (1981, p. 276), while Levi contends that a characterisation of chances independently of this principle is neither possible nor necessary (1980, p. 258). The principle itself, though a synthetic one, stands unsupported, and according to its advocates necessarily stands unsupported, by any empirical argument. The Kantian synthetic a priori has reappeared in the least likely of places.

Chances, it seems to follow, are simply postulated entities which justify numerically identical degrees of belief. That may well be all we know on Earth, but it is not all we need to know. Neo-Kantian metaphysics should not be taken as the foundation for a theory of objective probability.

■ e A BAYESIAN RECONSTRUCTION OF VON MISES'S THEORY

None of the theories we have looked at seems able to furnish an adequate, empirically significant, account of objective probabilities. We shall argue that this is not the case, however, and that von Mises's theory can, when, and only when, it is embedded in a theory of subjective probability. To this extent the relation between objective probabilities and observables is even less direct than that between the theoretical entities of physics and observables. There is, in particular, no deductive relation between the hypothesis that this experiment generates a collective and the outcomes of the first n tosses, for any n. Nevertheless, there is a relation between hypothesis and evidence, as we shall now show; but it is probabilistic and not deductive.

We shall proceed in stages. First, we need to establish a form of the Principal Principle. Suppose that you are asked to state your degree of belief in a toss of this coin landing heads, conditional upon the information *only* that were the tosses continued indefinitely, the outcomes would constitute a von Mises collective, with probability of heads equal to p. Suppose you answer by naming some number p' not equal to p. Then, according to the definition of degree of belief in Chapter 3, you believe that there would be no advantage to either side of a bet on heads at that toss at odds $\dfrac{p'}{(1 - p')}$. But that toss was specified only as a member of the collective characterised by its limit-value p. Hence you have implicitly committed yourself to the assertion that the fair odds on heads occurring at any such bet, conditional just on the same information that they are members of a collective with probability parameter p, are $\dfrac{p'}{(1 - p')}$. Suppose in n such tosses there were to be m at which a head occurred; then for fixed stake S (which may be positive or negative) the net gain in betting at those odds on heads would be $mS(1 - p') - (n - m)Sp' = nS\left(\dfrac{m}{n} - p'\right)$. Hence, since the limit of $\dfrac{m}{n}$ is p, and p differs from p', you can infer that the odds you have stated would lead to a loss (or gain) after finite time, and one which would continue thereafter. Thus you have in effect contradicted your own assumption that the odds $\dfrac{p'}{(1 - p')}$ are fair.

The Principal Principle, understood in this way, represents a consistent theory of single-case probabilities, and single-case probabilities which are equal to certain objective probabilities. But the single-case probabilities it introduces are not themselves objective. They are subjective probabilities, which considerations of consistency nevertheless dictate must be set equal to the objective probabilities just when all you know about the single case is that it is an instance of the relevant collective. Now this is in fact all that anybody ever wanted from a theory of single-case probabilities: they were to be equal to objective probabilities in just those conditions. The incoherent doctrine of objective single-case probabilities arose simply because people failed to mark the subtle distinction between the values of a probability being objectively based and the probability itself being an objective probability. Once the distinction is drawn, it becomes clear that single-case probabilities are really subjective probabilities which relative to the right sort of information are equal to probabilities in collectives.

We can now use the Principal Principle to give empirical significance to hypotheses about the values of von Mises probabilities, in the way sketched in our discussion of the theory of chance, namely by constructing posterior subjective-probability distributions over those hypotheses. Let, for example, h be the hypothesis that the outcomes of E are characterised by some given objective probability-distribution P, and let e state that the outcome of a performance of E is an instance of C. Let P now denote your subjective probability distribution. By Bayes's Theorem

$$P(h \mid e) = \frac{P(e \mid h)P(h)}{P(e)}.$$

$P(e \mid h)$ is computed according to the Principal Principle; $P(h)$ and $P(e)$ are the respective prior probabilities of h and e respectively. $P(h \mid e)$ gives your subjective probability of h in the light of the data e.

Sometimes merely a comparative verdict is all that is required, in which case $P(e)$ can be ignored. To take a simple, rather idealised, example, suppose that h and h' are two hypotheses stating that the tosses of this coin form a collective with probability p and p' respectively. Let e state that the coin was tossed n times, and that the observed frequency of heads was $\frac{r}{n}$. Then since the tosses constitute an initial segment of a collective, according to both h and h', it follows, as we noted earlier, that successive outcomes are independent with constant probability p and p' respectively. Relative to the collective determined by the n-fold repeated experiment $E_{(n)}$, therefore, the von Mises probability of the sample outcome is $p^r(1 - p)^{n-r}$ if h is true, and $p''(1-p')^{n-r}$ if h' is true. By the Principal Principle applied to these derived collectives, $P(e \mid h) = p^r(1 - p)^{n-r}$, and $P(e \mid h') = p''(1 - p')^{n-r}$, and

$$\frac{P(h \mid e)}{P(h' \mid e)} = \frac{p^r(1 - p)^{n-r}P(h)}{p''(1 - p')^{n-r}P(h')}$$

Suppose, now, that as n increases, the frequency $\frac{r}{n}$ approaches and remains within a very small neighbourhood of p'. Then, since the function of x, $x^r(1 - x)^{n-r}$, $0 < x < 1$, peaks very sharply, for large n, in the neighbourhood of $x = \frac{r}{n}$, and is close to 0 elsewhere, the ratio above will tend to 0, independently of the values of the prior probabilities $P(h)$ and $P(h')$, assuming neither is 0 or 1. In other words, the support of h by e will eventually be negligible compared with the support of h' by e.

We shall see in the next chapter how Bayes's Theorem enables us to evaluate many more types of statistical hypothesis in the light of sample data, in a way that is both very natural and which is vulnerable to none of the objections to the classical theories of statistical inference which we examined in chapters 5 to 8. The Bayesian theory will, moreover, turn out to provide the rationale for all those intuitively plausible prescriptions advanced by the classical, non-Bayesian theories, but which those theories themselves were unable to produce.

■ f ARE OBJECTIVE PROBABILITIES REDUNDANT?

Before we turn our attention to Bayesian statistical inference we must briefly consider the thesis, advanced by de Finetti and Savage, that the use of objective probabilities should be rigorously eschewed and that such probabilities are anyway redundant in just those contexts where they appear to possess explanatory value.

According to de Finetti and Savage, hypotheses about objective probabilities lack empirical significance (de Finetti, a strong positivist, actually declared them absolutely meaningless). It is sufficient to show them wrong in this claim actually to construct an empirically significant account of objective probability, and this we believe we have done. The redundancy thesis is not so easily disposed of, though it too is false, as we shall now show.

De Finetti proved the theorem, which he and Savage interpreted as implying the explanatory redundancy of objective probabilities, in (1937). Before stating this theorem and evaluating the de Finetti–Savage claim, some preliminary explanation is in order. Suppose that P is a subjective probability defined for all finite sequences of zeros and ones (which can mean 'having the property C', 'not having C' respectively, for any property C). Suppose also that P assigns a probability to every statement of the form 'the i_1th member of the sequence is x_1 and the i_2th member of the sequence is x_2 and ... and the i_kth member of the sequence is x_k', where each x_j is 1 or 0. Suppose, finally, that the P-probability of any such statement is the same whatever the order of the zeros and ones occurring in it. If P has these properties then the binomial random variables $X_1, \ldots, X_n, \ldots$, whose values on any sequence are the first, second, $\ldots$, nth, $\ldots$ members of that sequence, are called *exchangeable* (relative to P) by de Finetti.

What de Finetti then showed was that if the X_j are exchangeable, then the P-probability of any sequence of n of them taking the value 1 r times and 0 the remaining $(n - r)$ times, in some given order, is equal to

(1) $\displaystyle\int_0^1 p^r(1 - p)^{n-r}dF(p)$

where $F(p)$ is uniquely determined by the P values on the X_i, and has in addition the mathematical form of a distribution function taking the value 0 at $p = 0$ and 1 at $p = 1$, even though there is no random variable taking values p defined on the space of outcomes of the experiment.

But this last fact is very surprising, since (1) is exactly the value you would get if you believed (i) that there is a constant unknown probability of a 1 at any point in the sequence, where F defines your subjective probability-distribution over its values in the closed unit interval, (ii) that the X_i were independent relative to this unknown probability, and (iii) you evaluate the subjective probability of r ones and $(n - r)$ zeros to be equal to the value that that outcome would have relative to that unknown probability (you use the Principal Principle, in other words).

To see this, let h be the hypothesis that the X_i are independent with constant but unspecified probability, let h_p be the hypothesis that the value of this unknown probability is p, and let e be the statement that r ones are observed, in a sample of size n. If, for simplicity, we first assume that p is discrete and takes only finitely many values then, by the probability calculus,

$$P(e \mid h) = \sum_j P(e \mid h_{p_j} \,\&\, h)P(h_{p_j} \mid h)$$
$$= \sum_j p_j^r(1 - p_j)^{n-r}P(h_{p_j} \mid h)$$

In the passage from discrete to continuous p, the sum is replaced by an integral, $P(h_{p_j} \mid h)$ is replaced by the distribution function F, and we obtain an equation formally identical to (1).

Now it is customary to invoke the hypothesis of constant objective probability and independence (with respect to that probability)—in other words to invoke h—to explain those sorts of features of, for example, long sequences of coin tosses which von Mises singled out as requiring some sort of scientific explanation, namely the non-existence of gambling systems and the marked convergence of the relative frequencies. From our point of view the explanatory hypothesis is that a sequence of coin tosses is the initial segment of a von Mises collective (for it is in terms of collectives that we have elected to understand objective probabilities), which implies, as we saw, that the hypotheses of independence and constant probability are satisfied.

But de Finetti's proof implies that the P probability of any feature of the X_i obtained on the condition that h is true is obtainable simply by assuming that the X_i are exchangeable and without, or so it seems and de Finetti claims, introducing objective probabilities at all. Hence their introduction, at any rate from the point of view of explaining any given property of the X_i which we feel receives an explanation in terms of appeal to the hypothesis of independence plus constant probability from trial to trial, appears to be quite unnecessary.

■ g EXCHANGEABILITY AND THE EXISTENCE OF OBJECTIVE PROBABILITY

We shall now argue that de Finetti's thesis is untrue and that although objective probabilities are not explicitly introduced into the domain of P, they are nonetheless to a very great extent implicit in the very assumption of exchangeability itself. First, consider two events in the domain of P, (i) the sequence s_1, $010101 \ldots 01$, of length $2n$, and (ii) the sequence s_2, of the same length $2n$, also ending with 1, also having n zeros and n ones, but differing from s_1 in that the arrangement of its ones and zeros is very disorderly. Assume that the X_i are exchangeable, so that $P(s_1) = P(s_2)$. By the probability calculus

$$P(s_1) = P(1 \mid 010101 \ldots 0)P(010101 \ldots 0)$$

and

$$P(s_2) = P(1 \mid s_3)P(s_3),$$

where s_3 is the segment of s_2 of length $2n - 1$ preceding the terminal 1. By exchangeability again

$$P(010101 \ldots 0) = P(s_3)$$

and hence

$$P(1 \mid s_3) = P(1 \mid 010101 \ldots 0).$$

But these equations hold for all values of n, however large, and s_3 is by assumption a highly disorderly sequence in which there are as many zeros as ones. Were we actually to deny that the x_i are independent, there could be no reason for this last equation; we should in these circumstances expect the right-hand side to move to the neighbourhood of 1 and the left-hand side to tend to $\frac{1}{2}$ or thereabouts. We should certainly not expect equality. (This example is to be found in Good 1969, p. 21.)

This is not a proof that exchangeability implies independence (which is not true), but it is a very powerful reason for supposing that exchangeability assumptions would not be made relative to repeated trials of this type *unless* there was already a belief that the variables were independent. But there is in addition a proof, due to Spielman (1976), which we shall not reproduce here, that if the X_i are exchangeable, for all $n > 0$, then with P probability 1 the experiment will generate a von Mises collective. In other words, if you regard the X_i as exchangeable, then you must be certain that the relative frequency of ones tends to a limiting value and that whatever character is generated at the ith place in the sequence is independent, with respect to that limiting value of the characters generated both before and after that place (though that you are certain that the relative frequency of ones tends to a limit was in effect proved by de Finetti himself).

Spielman's important result requires that P be countably, and not merely finitely, additive, a condition that de Finetti remains strongly opposed to. But as we saw in Chapter 3, it is justified as a consistency constraint no less than the other axioms of the probability calculus, so we feel that Spielman is quite correct to disregard that objection. We shall, therefore, conclude that objective probabilities, far from being redundant, are on the contrary necessary to justify the exchangeability of the X_i.

■ h **CONCLUSION**

We have shown, admittedly only in outline so far, how hypotheses about objective probability distributions are to be evaluated against observational data. The only understanding of objective probability which permits empirical evaluations of hypotheses about its magnitude in given cases appears, moreover, to be in terms of long-run relative frequencies. So despite a very adverse press over the last quarter of a century, von Mises's theory seems, as he himself believed, to offer the only scientific account of objective probability, where by 'scientific' we mean that hypotheses about the magnitudes of such probabilities are in principle capable of empirical evaluation. The task now is actually to evaluate some particular types of hypothesis against some particular types of sample data and see what results are obtained. This will be the task of the next chapter.

CHAPTER 10

Bayesian Induction: Statistical Hypotheses

Bayesian induction involves the computation of a posterior probability, or density, distribution from a corresponding prior distribution, by means of Bayes's Theorem. This theorem does not discriminate between deterministic and statistical hypotheses; hence, unlike classical approaches, it affords a unified treatment for every kind of hypothesis. We have already considered, in Chapter 4, the way that Bayes's Theorem deals with a number of issues raised in the evaluation of deterministic hypotheses. In this chapter we shall, in the main, address the type of hypothesis that is most often discussed in expositions of classical statistical inference, namely, those characterised by the value of a parameter. We have already pointed out the necessity in the classical scheme of restricting possible hypotheses in such ways, and we have criticized the artificiality and arbitrariness of such restrictions. Our approach will be first to operate with the restricted models made standard by their widespread use in classical manuals, so as to allow a comparison. But in order to give a more realistic analysis, we shall also consider how and under what circumstances the restrictions might be lifted.

■ a THE PRIOR DISTRIBUTION AND THE QUESTION OF SUBJECTIVITY

In estimating the value of a parameter, the Bayesian approach, unlike the classical, starts out from a prior probability, or density, distribution over a set of possible values of the parameter. The distribution reflects a person's opinions before the results of the experiment are known. As we have made clear, such opinions are subjective, in the sense that they are shaped, in part, by elusive, idiosyncratic influences and that they may, and often do, vary from person to person. The subjectivity of the premisses of a Bayesian inference might suggest that the conclusion must be similarly idiosyncratic, subjective, and variable. Were this the case, the Bayesian approach would fly in the face of one of the most striking facts about science, namely its substantially objective character. However, it is not the case. As Edwards, Lindman, and Savage (1963, p. 527) have pointed out, "If observations are precise, in a certain sense, relative to the prior distribution on which they bear, then the form and properties of the prior distribution have negligible influence on the posterior distribution". Hence, from a practical point of view, "the untrammelled subjectivity of opinion about a parameter ceases to apply as soon as much data become available. More generally, two people with widely divergent prior opinions but reasonably open minds will be forced into arbitrarily close agreement about future observations by a sufficient amount of data". We may illustrate these points by reference to two sorts of problem that frequently crop up in statistical literature and practical research.

a.1 Estimating the Mean of a Normal Population

It will be recalled that we used this customary example to present the ideas of classical interval estimation. The population whose mean is to be estimated is assumed to be normal and to have a known standard deviation, σ. The assumption of a known standard deviation is a plausible approximation in many cases, for instance, where an instrument is used to measure some physical quantity. The instrument would, as a rule, deliver a spread of results if used repeatedly under similar conditions, and experience shows that this variability often follows a normal curve, with an approximately constant standard deviation. Making measurements with such an instrument would then be practically equivalent to drawing a random sample of observations from a normal population of possible observations, whose mean is the unknown quantity and whose standard deviation has been established from previous calibrations.

Let θ be a variable that ranges over possible values of the population mean. We shall assume for the present that the prior density distribution over θ is also normal, having a mean of μ_0 and a standard deviation of σ_0. In virtually no real case would the prior distribution be strictly normal, for physical considerations would place limits on possible values of a parameter, while normal distributions assign positive probabilities to every range. For instance, the average height of a human population could not be negative, nor could it be five thousand miles. Nevertheless, a normal distribution often provides a mathematically convenient idealisation of sufficient accuracy. (We are assuming a normal prior distribution in order to simplify this illustration of Bayesian inference at work. We shall show later that the assumption may be considerably relaxed without affecting the conclusions substantially.)

Let a random sample of size n be drawn from the population, and let $\bar{x}$ be the mean of that sample. It is now, surprisingly, a simple matter to calculate the posterior distribution of θ, relative to these data. This distribution is also normal (like the prior distribution) and its mean, μ_n, and standard deviation, σ_n, are given by

$$\mu_n = \frac{n\bar{x}\sigma^{-2} + \mu_o\sigma_o^{-2}}{n\sigma^{-2} + \sigma_o^{-2}} \text{ and } \frac{1}{\sigma_n^2} = \frac{n}{\sigma^2} + \frac{1}{\sigma_o^2}.$$

These results (which are proved by Lindley, 1965, vol. 2, for example) are illustrated below.

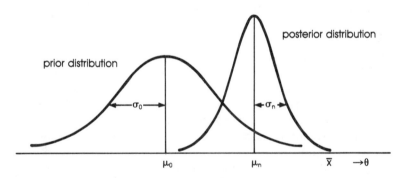

The prior and posterior distributions for the mean, θ, of a population with mean and standard deviation μ and σ, relative to a sample with mean $\bar{x}$.

The precision of a distribution is often defined as the reciprocal of its variance. So the second of the equations above tells us that the precision of a posterior distribution increases with the precision σ^{-2} of the population whose mean is being evaluated. By the same token, the precision of a measuring instrument is the reciprocal of the variance of its error distribution. Thus the more precise such an instrument is, the more precise will be the corresponding posterior distribution of the quantity being measured. Assuming that a precise inference is desirable, this accounts for the appeal of the efficiency criterion that classical statisticians have imposed, ad hoc, on estimators. (*See* Chapter 8, section **b.4.**)

It is easy to see, too, from the above formulas that, as n increases, μ_n (the mean of the posterior distribution) tends to $\bar{x}$ (the mean of the sample). There is a similar result for the variance, σ_n^2 of the posterior distribution, namely that it tends to $\frac{\sigma^2}{n}$, a quantity that is independent of the prior distribution, being fixed just by the population (through its standard deviation, σ) and the size of the sample. Thus, as the sample is enlarged, the contribution of the prior distribution and, so, of the subjective part of the inference, lessens, eventually dwindling to insignificance. Moreover—and this is the crucial point for explaining the objectivity of the inference—the objective information contained in the sample becomes the dominant factor relatively quickly. Hence, two people proceeding from different normal prior distributions would, with sufficient data, arrive at posterior distributions that were arbitrarily close together and, moreover, this convergence of opinion would be rather rapid.

We can show how rapid the convergence may be by an example in which one person's normal prior distribution over the mean of some normal population is centred on 10, with a standard deviation of 10, and a second person's is centred on 100, with a standard deviation of 20. This difference represents a very sharp division of initial opinion, for the region that each considers almost certainly contains the true mean is regarded by the other as practically certain not to contain it. But even profound disagreements such as these are resolved after relatively few observations. The following table shows the means and standard deviations of the posterior distributions for the two people, relative to random samples of various sizes, each sample having a mean of 50. We shall assume that the population is normal and has a standard deviation of 10.

We have selected a rather extreme example, in which the two people differ substantially in their prior opinions, there being almost no overlap in the initial distributions. The first line in the above table, corresponding to no data, represents this initial situation. Nevertheless, a sample of only 20 brings the two posterior distributions very close, while one of 100 renders them indistinguishable. Not surprisingly, the closer opinions are initially, the less evidence is needed to bring the corresponding posterior opinions within given bounds of similarity. (This is proved, for instance, by Pratt et al., 1965, Chapter 11.) Hence, although a Bayesian analysis of the case under consideration must operate from a largely subjective prior distribution, the most powerful influence on its conclusion is the experimental evidence. We shall see that the same is true of more general cases.

TABLE XI The means and standard deviations of the posterior distributions of 2 people with different prior distributions, relative to a random sample with a mean of 50. The population standard deviation is 10.

Sample size	Person 1		Person 2	
n	μ_n	σ_n	μ_n	σ_n
0	10	10	100	20
1	30	7.1	60	8.9
5	43	4.1	52	4.4
10	46	3.0	51	3.1
20	48	2.2	51	2.2
100	50	1.0	50	1.0

a.2 Estimating a Binomial Proportion

Consider another standard problem in statistics, namely, the estimation of a proportion, say, of counters in an urn, or of a physical probability, say of a coin turning up heads or of a person recovering from an illness. Data are collected by sampling a population at random or by performing an appropriate experiment, like flipping the coin. If such trials have just two possible outcomes, with probabilities p and $1 - p$, each of which is constant from trial to trial, then the data-generating process is known as a Bernoulli process, and p and $1 - p$ are called the Bernoulli parameters or binomial proportions. The two outcomes of a Bernoulli process are often labelled 'success' and 'failure'.

The Bayesian approach to the estimation of p commences with the description of a prior distribution. If we assume that this takes the form of *a beta distribution,* the problem is very simply solved. (*See,* for example, Pollard, Chapter 8.) (Again, we shall later show that this assumption may be very considerably weakened without substantially affecting the conclusions.) A beta distribution, like a normal one, embodies a family of distributions whose particular shape is fixed by two parameters, u and v. A beta distribution may have a wide variety of shapes, depending on u and v, as the diagram below shows. For instance, if $u = v = 1$, the distribution is uniform. The mean and variance of a beta distribution is given by

$$\text{mean} = \frac{u}{u + v} \qquad \text{variance} = \frac{\left(\dfrac{u}{u + v}\right)\left(\dfrac{v}{u + v}\right)}{u + v + 1}$$

SOME BETA DISTRIBUTIONS

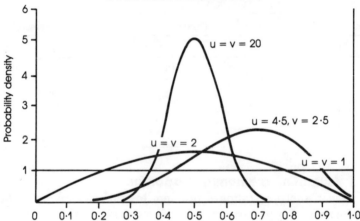

If the prior distribution over the Bernoulli parameters is of the beta form, Bayes's Theorem is particularly easy to apply. Suppose a random sample of n observations obtained from a Bernoulli process shows s successes and f failures. It turns out that the posterior distribution is also of the beta form, with parameters $u' = u + s$ and $v' = v + f$. Hence, the mean of the posterior distribution is $\dfrac{u + s}{u + v + n}$, which tends to $\dfrac{s}{n}$, as the number of trials increases to infinity, while the variance of the posterior distribution tends, though more slowly, to zero. Thus, as with the earlier example, the influence of the prior distribution upon the posterior distribution steadily diminishes with the size of the sample, the rate of diminution being considerable, as simple examples, which the reader may construct, would show.

■ b CREDIBLE INTERVALS AND CONFIDENCE INTERVALS

Estimates of parameters are often reported in the form of a range of possible values, e.g., $\theta = \theta^* \pm \epsilon$. Such estimates have an obvious Bayesian interpretation, namely, as a range that, with very high probability, contains the true value of θ. This would be an example of a so-called credible interval. Such intervals may be calculated from a posterior distribution. If P is the probability that θ lies between a and b, then the interval (a,b) is said to be a $100 \times P$ per cent credible interval for θ. Bayesians recommend credible intervals as useful summaries of posterior distributions.

The idea may be illustrated with the first example we gave. As we showed, in the presence of sufficient data, the posterior distribution of the mean of a normal population is symmetrical about the sample mean, $\bar{x}$, and has a standard deviation $\sigma_n = \sigma n^{-\frac{1}{2}}$, where n is the sample size, and σ is the population standard deviation. Since the distribution is normal, the range $\bar{x} \pm 1.96 \times \sigma n^{-\frac{1}{2}}$ contains θ with a probability of 0.95 (*see* Chapter 2, section **f.3**), and so is a 95 per cent credible interval.

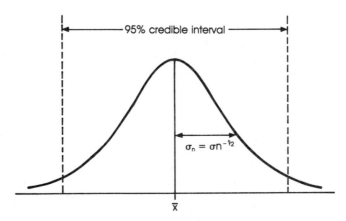

A posterior distribution showing a 95% credible interval for the mean of a normal population with standard deviation σ

Of course, by selecting other areas of the distribution, one could construct different credible intervals. For instance, the infinite range of values defined by $\theta > \bar{x} - 1.64 \times \sigma n^{-\frac{1}{2}}$ is also a 95 per cent credible interval. Another would be an interval that extends further into the tails of the posterior distribution but omits a narrow band of values around the centre of the distribution. There is sometimes a discussion about which of the possible intervals should be 'chosen', which we believe is misconceived (*see,* for example, Lindley, 1965, vol. 2, pp. 24–25), for strictly speaking, one should not *choose* any interval, because a choice implies a commitment in excess of that permitted by the probability of the interval. All 95 per cent credible intervals are on a par.

Credible intervals clearly resemble the confidence intervals by which classical statisticians estimate parameters. Indeed, in the particular case before us, the 95 per cent credible interval depicted in the diagram and the 95 per cent confidence interval that is routinely favoured (i.e., the shortest one) coincide (*see* Chapter 8, section **c.2.iv**). But although credible intervals and confidence intervals are similar, they differ in a crucial way. In particular, confidence interval estimates do not state the probability relative to the evidence that θ lies within the interval (classical statisticians say this is meaningless); they express the probability that confidence bounds measured in a specified way lie on either side of θ, given that θ is the true value of the parameter. Bayes's Theorem, of course, permits the calculation of the former from the latter and so justifies a conclusion about how confident one should be that θ is confined within some interval. No such conclusion is allowed by the classical scheme, even though, from an intuitive point of view, it is irresistible. Classical statisticians follow intuition insofar as they often claim that a confidence coefficient is a measure of the degree of confidence appropriate to the belief that the corresponding interval contains the true value of the parameter; indeed, this interpretation is signposted in the name. But, as we argued, such an interpretation of confidence intervals is illegitimate within the constraints of classical statistics. Moreover, it is hard to know what a degree of confidence could amount to when one is restricted to purely objective appraisals as classical statisticians supposedly are. Bayesian credible intervals, on the other hand, provide an intelligible interpretation and a rational explanation for the intuition that appears to underlie classical confidence intervals.

■ c THE PRINCIPLE OF STABLE ESTIMATION

The inferences in the two examples we have given are very insensitive to variations in the prior distributions, as we said. However, we restricted the argument to cases of, in the first example, normal, and in the second, beta, distributions and the question arises whether the same insensitivity would exist if these restrictions were relaxed. A theorem known as the Principle of Stable Estimation, due to Edwards, Lindman, and Savage (1963), shows that even a very considerable relaxation would preserve that insensitivity. We shall indicate the scope of the theorem. It states that if the prior distribution satisfies certain conditions, which it specifies, the posterior distribution is approximately the same, to an extent that it specifies, as it would have been if the prior were uniformly distributed. Hence, provided the prior beliefs of different people meet the conditions of the theorem, the corresponding posterior beliefs at which each would arrive would be roughly the same.

In order to ascertain whether the conditions of the theorem apply, one must go through three steps. First, a 99 per cent or higher (the higher, the better) credible interval, covering some range B, should be calculated on the hypothesis that the prior distribution is uniform. Since a uniform distribution is only defined over an interval that is bounded at both ends, this will involve setting limits to the values that the parameter could have. Secondly, one must ascertain that within B, the actual prior is almost constant; more specifically, the range of probability values within B should not be more than about 5 per cent (but the lower, the better) of the minimum probability value in B. Thirdly, one must verify that most (say 95 per cent, but the more the better) of the posterior distribution corresponding to the actual prior is covered by B. As the authors pointed out, the third condition may be difficult to check in practice. For practical purposes, they proposed an alternative, which is actually stronger than the third condition, but which would normally be easier to verify. This is that nowhere outside B should the actual prior be astronomically big compared with its value within B (the authors note that a factor of up to 1000 would be tolerable). The Principle of Stable Estimation asserts that if the three conditions are met, the actual posterior distribution and that derived on the basis of a uniform prior are approximately the same. We shall omit the details of the approximation, save to point out that the better the conditions are met, the more satisfactory the approximation, and that in general a larger sample will satisfy the conditions better than a smaller one. The conditions under which the Principle of Stable Estimation applies are neatly illustrated by the following diagram, taken from Phillips (1973):

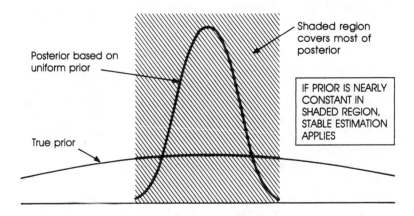

Practical application of the Principle of Stable Estimation

The Principle of Stable Estimation ensures that the relative insensitivity to variations in the prior distribution that we noted in the examples above is not confined to distributions belonging to any particular family. It thus facilitates a more general response to the fear that Bayesian inference is unacceptably subjective. The Principle also defuses the difficulty that would arise in applying Bayes's Theorem to parameter estimation if, as might at first appear, one had to start with an accurate and precise description of someone's prior distribution. For although people can often compare the probabilities of their beliefs very roughly, they find it difficult to assign precise values to those probabilities. The Principle shows that provided the sample is sufficiently large, no great precision is required when describing the prior distribution in order to arrive at a correct and relatively precise posterior distribution.

■ d DESCRIBING THE EVIDENCE

The result of an experiment is a physical phenomenon or state; on the other hand, evidence that is used in the appraisal of a hypothesis must be expressed as a statement. The question therefore arises of which aspects of the physical situation the evidence should incorporate. Clearly not every aspect could be described, nor would anyone think a complete description desirable as an ideal, for some aspects of an experimental result are plainly irrelevant. For instance, the colour of the experimenter's socks when sowing plant seeds would, presumably, not be worth recording if one were interested in the genetic structure of the plant. Clearly, evidence need not describe irrelevant facts such as these. On the other hand, it should omit no relevant information.

What constitutes relevance in this context and how can one tell what is relevant and what is not? The answer to the first question is that a fact is relevant to a set of hypotheses when a knowledge of it makes a difference to one's appraisal of those hypotheses; in relation to the Bayesian inductive theory, this means that the fact influences the probabilities of the hypotheses. The factors determining whether a fact is relevant in a particular case may be illustrated by reference to the type of hypothesis we have been considering in this chapter, namely, those that comprise an exhaustive and mutually exclusive set of hypotheses, each of which is characterised by a particular value of some parameter, say θ. For simplicity, we shall relate the discussion to the example of an experiment designed to elicit the physical probabilities of a coin to land heads and tails. We regard this as an elementary representative of a type of experiment that crops up frequently in science. Suppose the coin has been tossed several times and let us consider a number of descriptions of the result, each description being accurate, but varying from others in its detail.

The first description, e_1, contains the least information. Intuitively, its evidential value is rather small, since it says nothing about n, the number of times the coin was tossed. A Bayesian inference from the result would require not only a prior distribution over θ (the physical probability of the coin to land heads), but one over n as well. It is not necessary for us to enter further into this rather complicated case, save to note that unlike the classical approach, and more plausibly, we think, the Bayesian does not necessarily find e_1 entirely uninformative.

TABLE XII Possible descriptions of the result 4 heads and 6 tails, obtained after tossing a coin 10 times

---

e_1: 4 heads.

e_2: 4 heads, 6 tails in a trial designed to have 10 throws

e_3: 4 heads, 6 tails in a trial designed to cease after 6 tails

e_4: the sequence HHTHTTTHTT

e_5: the sequence HHTHTTTHTT obtained in a trial designed to terminate when the experimenter is called for lunch

e_6: 4 heads, 6 tails

e_7: the sequence HHTHTTTHTT, with the first throw being performed with the left hand, the second . . . and the sixth with the right hand

e_8: The experimenter intended to report the outcome either as the occurrence of 4 heads and 6 tails, or as its non-occurrence. It did occur.

Let us proceed to e_2, which states n exactly and, moreover, gives the stopping rule which an experimenter used. This enables the posterior probability distribution to be calculated in accordance with Bayes's Theorem, $P(\theta \mid e) = \dfrac{P(e \mid \theta)P(\theta)}{P(e)}$. In the present instance, $e = e_2$, $P(e \mid \theta) = {}^nC_r\theta^r(1 - \theta)^{n-r}$, and $P(e) = \int {}^nC_r\theta^r(1 - \theta)^{n-r}P(\theta)d(\theta)$, where $n = 10$ and r (the number of heads) $= 4$. The binomial factor, ${}^nC_r$, being a function of n and r only, is independent of θ, and so it cancels from top and bottom in Bayes's Theorem. Clearly any other description of the trial for which $P(e \mid \theta) = K\theta^r(1 - \theta)^{n-r}$, where K is independent of θ, yields the same posterior distribution. This would be the case with e_3, which states a different stopping rule, and e_4, which describes the precise sequence of heads and tails in the outcome. In the former case, $K = {}^{n-1}C_{r-1}$ (*see* Chapter 7, section **c.5**), and in the latter, $K = 1$. Hence, in the Bayesian approach, it does not matter whether the experimenter intended to stop after n tosses of the coin or after r heads appeared in a sample; the inference about θ is exactly the same in both cases. As we have already noted, this is not so in the classical scheme.

e_5 is a case where the rule to stop the trial depends on some external event rather than on any feature of the sample. e_5 is equivalent to the conjunction e_4 & 1, where 1 states that lunch was ready just after the tenth throw of the coin, and, as before, e_4 describes the sequence of heads and tails in the outcome. In applying Bayes's Theorem, we need to consider the quantities $P(e_4$ & $1 \mid \theta)$, which can also be expressed as $P(e_4 \mid \theta)P(1 \mid e_4$ & $\theta)$. If the point at which lunch is ready is independent of θ, as presumably it is, then $P(e_5 \mid \theta)$ equals $P(e_4 \mid \theta)P(1 \mid e_4)$, hence it is of the form $KP(e_4 \mid \theta)$, where K is the same for all θ. Then, as before, K cancels from Bayes's Theorem and so e_5 results in the same posterior probabilities as e_4. Therefore, as one would expect, this stopping rule has no inductive relevance.

Data are normally reported without mention of the stopping rule; for instance, e_6 merely states that the outcome contained 4 heads and 6 tails. But this poses a problem, for the probability of such a result depends not only on the parameter, but also on the manner in which the trial was conducted, in particular in regard to the stopping rule. Hence, without more information, one could not compute $P(e_6 \mid \theta)$, so that neither Bayes's Theorem nor even a classical test of significance could be applied. The difficulty is usually met by simply assuming that the sample size was fixed in advance, whether or not there is evidence that this was so, and then evaluating the probability of the evidence relative to θ accordingly. This arbitrary-seeming practice can be justified. The argument is this. Continuing with the coin example, we know that the 4 heads and 6 tails in the sample must have occurred in some order. Whatever that order was, its probability conditional on θ is $\theta^4(1 - \theta)^6$, so we may use this as the appropriate input into Bayes's Theorem. But we could also have used the term $K\theta^4(1 - \theta)^6$, without affecting the resulting posterior distribution, provided K is the same for all θ. Hence we are permitted to assume that the sample was drawn with a stopping rule that fixed its size in advance, for this is merely equivalent to assuming that K is a constant equal to $^{10}C_4$.

The stopping rule may, in certain circumstances, contain relevant information, however. This would be so if one were relying upon a random sample to measure the mean height of a group of cooks who happened to be preparing lunch at the same time as the experiment was in progress. Suppose tall chefs cooked faster than short ones and the trial was concluded as soon as lunch was ready. In this instance, exactly when the trial was concluded would depend on the unknown parameter and so would be some guide to the stature of the cooks. Ignoring the stopping rule in such a case would be overlooking relevant information. Hence, the stopping rule is not necessarily uninformative, but as this example suggests, normally it would be. This concession should not be misunderstood. It does not mean that the scientist's *intention* to stop the trial at a particular point is of any inductive significance, an interpretation that would lead to contradictions if a number of scientists worked on the same trial but with a variety of intentions. Hence, our position is quite different from that of the classical statistician. We are simply claiming that in estimating a parameter, one normally would derive all one's information from the composition of a random sample, but that sometimes events occurring during the sampling process also have some significance as evidence.

The description e_7 in the list of possible descriptions of the coin-tossing trial seems to contain irrelevant information, an intuition endorsed by Bayes's Theorem. For e_7 is equivalent to the conjunction e_4 & h, where e_4 describes the sequence of heads and tails and h states which hand was involved in each throw of the coin. As with an earlier example, if h is independent of θ (which it usually would be), it is irrelevant to θ. Similar considerations apply to the last example, e_8. Presumably an experimenter's intention to ignore certain aspects of results which did not in fact eventuate would have no psychokinetic effect on the experiment and so would not affect the probability of the actual result. If so, information about that intention is not relevant and has no evidential significance in this context.

These conclusions, it seems to us, are unequivocally endorsed by most people's intuitions. It will be recalled, however, that classical tests and estimation procedures contradict them. Despite the classical approach having dominated the scene for above sixty years, these consequences of it still seem surprising and unacceptable.

■ e SUFFICIENT STATISTICS

In an experiment to determine the parameters of what is believed to be a Bernoulli process, one would normally just count up the number of successes and the number of failures in a random sample and ignore the order in which they occurred. Similarly, the mean of some population would normally be estimated by the mean of an appropriate sample, the specific details of each measurement being ignored. However, such details are not invariably superfluous. If, for instance, a thousand throws of a coin produced an equal number of heads and tails, it would surely be useful to know that all the heads preceded all the tails, for the extra information might point to some change in the physical character of the coin which one would not have guessed from the summary data. The question whether the evidence should include a description of each measurement and its order of presentation in the sample, or whether some of the information may be ignored, is often dealt with through the theory of sufficiency.

It will be recalled that the classical theory of estimation and Fisher's version of significance tests assert that the statistics employed for an inference ought to be sufficient, on the grounds that only sufficient statistics contain all the information that is relevant. We have already pointed out that the notion of relevancy to which this reason appeals is an intuitive one that has to be grafted onto classical views of statistical inference. The position is different for the Bayesian theory, where the natural notion of relevancy (a datum is relevant if and only if it makes a difference to the final assessment of a hypothesis) automatically implies the inductive sufficiency of statistics that are sufficient in the technical sense. We showed this earlier (in Chapter 8) with the demonstration that $p(\theta \mid t) = p(\theta \mid \mathbf{x})$ if, and only if, t is sufficient relative to some datum $\mathbf{x} = x_1, \ldots, x_n$ and in the context of the set of hypotheses represented by the possible values of θ. That is, the posterior probability distributions conditional on t and on $\mathbf{x}$ are the same.

This conclusion reminds us that sufficiency is defined with respect, first, to a set of data and, secondly, to a set of hypotheses. In the Bayesian case, these hypotheses have to exhaust all the possibilities, in the sense that $\Sigma P(\theta_i) = 1$. Of course, a statistic might be regarded as being 'almost sufficient' and so practically adequate, if $\Sigma P(\theta_i)$ were somewhat less than 1. It would be practically adequate in the sense that the posterior distributions relative to $\mathbf{x}$ and to t would be sufficiently similar for a given purpose (Pratt, 1965, pp. 169–171). The statistic would be inadequate, however, if some moderately plausible hypothesis, call it h, that is not represented by a numerical value of θ, implied that $P(\mathbf{x} \mid t \,\&\, h)$ is moderately high and that it differed substantially from $P(\mathbf{x} \mid t)$. For in that case, $P(h \mid \mathbf{x})$ would depart substantially from $P(h \mid t)$. For instance, if t described the proportion of successes in a Bernoulli trial, and $\mathbf{x}$ the sequence of successes and failures, h might attribute to the trial some variable probability of producing successes, or it might accuse the experimenter of cheating in some way. This explains the fact, which we noted earlier, that the order of a sequence of observations may sometimes convey important information.

The classical account of sufficiency is similar to the Bayesian one. However, it differs in two notable ways, both deriving from its failure, except in rare cases, to treat hypotheses as entities that may be qualified by a probability. First, according to the classical definition, t is sufficient relative to $\mathbf{x} = x_1, \ldots, x_n$ if $P(t \mid \mathbf{x})$ is independent of θ, where different values of θ represent different hypotheses. This resembles the Bayesian definition, but unlike it, θ does not appear in the conditional probability term, since it is not regarded as a random variable by classical statisticians. Secondly, and more importantly, the set of hypotheses relative to which sufficiency is defined has to be circumscribed in an arbitrary way. Clearly, it ought to be limited, because if it contained every logical possibility, then it would, for instance, include a hypothesis (indeed, infinitely many hypotheses) that predicted with certainty the particular sequence $x_1, \ldots, x_n = \mathbf{x}$; hence no statistic based on $\mathbf{x}$ (except, trivially, one that retained all the information in $\mathbf{x}$) would ever be sufficient. On the other hand, the hypotheses relative to which sufficiency is defined cannot be selected on a probability criterion, for this would concede the case to the Bayesians. There appear to be only two other possibilities: classical statisticians may envisage some yet-to-be-described, non-probabilistic means by which hypotheses could be judged worthy of consideration or the hypotheses could be picked in an arbitrary way, say, by tossing a coin or by a blind selection from a hatful of hypotheses. Neither seems equal to the situation. The second is hardly in keeping with what we understand by the scientific attitude; the first is difficult to make specific, particularly when one bears in mind the guiding principle of classical reasoning, namely, that inductive inferences should be purely objective.

■ f TESTING CAUSAL HYPOTHESES

We discussed how to test causal hypotheses earlier, in the context of Fisher's theory because Fisher created the methodology of such tests which is now so widely followed. It will be recalled that the novelty of Fisher's approach was to require a process known as randomization. In Chapter 6 we argued that although it is often acknowledged to be a prerequisite, the randomizing of treatments in a trial does not do the job required of it. Randomization was intended to allow rational inferences to be drawn from agricultural, medical, and similar trials without prejudicing the supposed objectivity of science, the particular problem being that when two treatments produce different effects in different circumstances, any of innumerably many factors might be responsible. But as we have explained, the aim could not be achieved by Fisher's methods, advocates of the randomization technique being forced to appeal to personal judgments, that is, to subjective assessments of the plausibility of various theories. We shall present here the outline of an alternative and, we believe, intuitively more satisfactory solution to the problem of evaluating causal hypotheses (*see* Urbach 1985 and 1987a).

Consider the matter through a medical example. Suppose a new drug were discovered which, because of its structural similarity to a chemical produced in humans who suffer from and subsequently overcome depression, seems to be an effective treatment for that condition. A clinical trial to test this idea would take the following form. Two groups of sufferers would be constituted. One of them, the test group, would receive the drug while the other, the control group, would not. In practice the experiment would be a little more complicated than this; for example, the people in both groups would be led to believe they were receiving an effective treatment. This can be achieved by the use of a placebo within the control group, that is, a substance which seems to the patient like the one being tested but which has no relevant pharmacological activity. Moreover, a fastidiously conducted trial would ensure that the doctor does not know whether he or she is administering the drug or the placebo (this restriction is known as a double blind). And one would also ensure that any other factors thought likely to influence recovery were present to an equal degree in each of the groups. That is, these factors are controlled.

Why such a complicated experiment? Well, the reason for a control group is obvious. One is interested in the causal effect of the drug on the chance of recovery; so ideally one wants to compare how patients responded with the drug with how they would have responded without it. The conditions in the control group are intended to simulate the latter circumstance. Now let e describe the results of the trial in terms of the relative recovery rates in each group, and let h be the hypothesis attributing the difference between responses of the two groups to the drug. Ideally the experiment would allow one to attribute any difference in recovery rates directly to the drug—or in Bayesian terms, the experiment would maximize the posterior probability of h relative to e. And Bayes's Theorem tells us that this aim can be achieved only if the trial is adequately controlled. According to that theorem, when $P(e \mid h)$ is fixed, the smaller $P(e \mid {\sim}h)$ is, the greater is the posterior probability of h; so in designing the trial, one would want $P(e \mid {\sim}h)$ to be as small as possible. In other words, e provides the strongest evidence for h when it would be maximally probable if h were true and minimally probable if h were false. Now $P(e \mid {\sim}h)$ is equal to the sum $\Sigma P(e \mid h_i)P(h_i)$, where $\{h_1, h_2, \ldots, h_n\}$ is an exhaustive set of mutually exclusive hypotheses whose disjunction is equivalent to ${\sim}h$. Hence, if among the hypotheses $h_1, \ldots, h_n$ there were some moderately plausible hypotheses, which explained the evidence reasonably well (by conferring a reasonably high probability on e), then $P(h \mid e)$ would be correspondingly lower.

Suppose no placebo control had been used; in that case, there would be at least one such hypothesis. For in the absence of a placebo control, it would be moderately likely that the difference in recovery rates in the two groups was due to a higher number of people in the test group confidently expecting to recover because they were receiving medical attention (call this hypothesis h_p). If, on the other hand, a placebo control were imposed, by subjecting both groups to ostensibly similar kinds of medical attention, then it would be much less likely that the people in the two groups had different expectations of recovery. The placebo control thus diminishes $P(h_p)$ and by so doing, it diminishes $P(e \mid {\sim}h)$; hence, $P(h \mid e)$ is increased. This is why a placebo control would be used.

Some controls are not worth applying. For instance, there would be little point in equalizing the two groups with respect to the length of the socks or the Christian names of the grandmothers of the medical personnel or the colour of the painted undercoat on the surgery walls, for the corresponding hypotheses that attribute a causal influence to such factors are already very improbable indeed. To diminish these probabilities further by imposing controls would be a waste of effort.

In this approach there is no general need to allot patients to the test and control groups in a random fashion and, of course, similar conclusions apply to agricultural, psychological, and other such trials. Random allocation may nevertheless sometimes be useful and appropriate. For instance, it is one way of mitigating the unconscious tendency that doctors often have to pick the less ill patients to receive a new treatment in a trial. However, allocating patients with the help of a physical stochastic process is clearly not the only way of dealing with such biases. One might, for example, get some independent, medically ignorant person to make the allocation, preferably without inspecting the patients. Best of all would be to make a careful study of the factors influencing the course of an illness and to balance the test and control groups accordingly. In other words, any method of allocation is satisfactory so long as there is no reason to think that it will prejudice the outcome of the experiment.

Summary. Bayes's Theorem supplies coherent and intuitively satisfactory guidelines for the design of clinical and similar trials. This contrasts with the classical principles governing such trials. One striking difference between the two approaches is that the second simply takes the need for controls for granted, while the first explains that need and, moreover, distinguishes in a plausible way between factors that have to be controlled and those which do not. Another difference is that the Bayesian approach does not make the random allocation of subjects to treatments a universal requirement. We regard this as a considerable merit, since we have discovered no good reason for regarding a random allocation as an indispensable precondition and several good reasons for not so regarding it (*see* Chapter 6).

■ g CONCLUSION

The Bayesian way of estimating parameters allows one to associate different degrees of confidence with different values and ranges of values, as classical statisticians sought to do, but it does this in a simple and straightforward way through its central principle and without the need for new conditions and the consequent need for new justifications. Hence, the Bayesian treatment of statistical and deterministic theories is the same and is backed by the same philosophical idea. The Bayesian also accounts for the intuitive sufficiency-condition and for the natural preference for maximum precision in estimators. It does not find a place for the condition of bias or that of consistency, but as we have argued, these criteria have nothing to recommend them anyway.

The Bayesian method also avoids one of the most perverse features of classical methods, namely a dependence upon the stopping rule, and its theory of clinical trials is, in our view, both coherent and intuitively right. There is a subjective element in the Bayesian approach which offends many, but this element, we submit, is wholly realistic; perfectly sane scientists with access to the same information often do evaluate theories differently, although, as the Bayesian predicts, they normally approach a common view as the evidence accumulates. The Bayesian also anticipates the possibility of people whose predispositions either for or against certain theories are so pronounced and different from the norm that their opinions remain eccentric, even with a large volume of relevant data. You might take the view that such eccentrics are pathological, that each theory has a single value relative to a given body of knowledge, and that responsible scientists ought not to allow personal, subjective, factors to influence their judgment. But then you would have to show why this should be the case and how it is possible. Many have attempted this, but none has succeeded.

■ PART V

Finale

Our next, and last, chapter is also one of the longest in this book, fittingly enough, since it sets out the objections which have been raised against the theory of method we have advanced in the previous chapters, and those objections are many and various. Without more ado, we shall pass straightaway to their consideration and—we believe—rebuttal.

CHAPTER 11

Objections to the Subjective Bayesian Theory

■ a INTRODUCTION

In the preceding chapters we have developed the theory of subjective or personalistic Bayesianism as a theory of inductive inference. We have shown that it offers a highly satisfactory explanation of standard methodological lore in the domains of both statistical and deterministic science; and we have also argued at length that all the alternative accounts of inductive inference—like Popper's or Fisher's—achieve their explanatory goals, where they achieve them at all, only at the cost of quite arbitrary stipulations. However, the subjective Bayesian theory itself has been the object of much critical attention, to such an extent, in fact, that it is still regarded in some influential quarters as vitiated by hopeless difficulties. These difficulties, in our view, stem from misunderstanding and confusion, and in this final chapter we shall do our best to dispel both.

Of the standard criticisms some—due largely to Popper and his followers—are answered relatively simply and quickly, and we shall deal with these first. We shall then consider an objection, due to Clark Glymour, which points to an apparently insuperable problem in explaining within any Bayesian theory how a hypothesis can be supported by data already known at the time the hypothesis was proposed. We often do want to say that hypotheses may be so supported: it is, after all, something of a commonplace that Einstein's General Theory of Relativity was supported by the value already accepted at the time of the seconds of arc through which Mercury's perihelion annually precesses; and the reader can no doubt think of many other examples. Glymour's objection, however, is precisely that the Bayesian theory is incapable of explaining how any data can support a theory proposed after the data became known.

We shall argue that Glymour's objection is false: the Bayesian theory can explain how data already known can support theories. However, the successful rebuttal of Glymour's objection appears to bring another in its train. This is that the Bayesian is incapable of discriminating, in his assessment of the support of a hypothesis by evidence, between evidence obtained independently of that hypothesis and evidence the hypothesis was deliberately constructed to explain. Then, runs the objection, and it seems prima facie a very powerful one, the Bayesian theory must be incorrect since quite clearly there should be no support of the hypothesis by the evidence in the second case. We shall show that despite its apparent plausibility, this last claim is incorrect, and that on the contrary there are many well-known examples of scientific theories drawing support from data which they were constructed to explain. Moreover, it turns out that the Bayesian approach reproduces exactly the sorts of informal reasoning actually employed in cases like these.

Though they come last, the remaining three objections have been the most influential. One, which has dominated the discussion for practically the whole of the present century, is that a subjective, degree-of-belief interpretation of the probabilities in Bayes's Theorem is inadequate precisely because it would make science a purely subjective affair. How then, it is objected, can the subjectivist explain the widespread agreement that science is correctly opposed to superstition in its claims on our credence because and only because it is based on objectively justifiable canons of inference, not on what people, perhaps whimsically, actually do believe and the extent to which they believe it? We shall argue that this objection rests on a confusion, and that a Bayesian reconstruction of the procedures of inductive inference poses no threat whatever to the objectivity of the scientific enterprise.

The next objection concerns the principle of conditionalisation, that is to say the principle that if $P(h \mid e)$ is your conditional probability of h on e, and you learn e (but nothing stronger), then consequent upon this information, your degree of belief in h is, if you are consistent, equal to $P(h \mid e)$. A frequently made charge is that the use of this principle commits the Bayesian to the unrealistic and certainly unwelcome existence of some Ur-distribution, from which his current distribution of belief is obtained by successive applications of the principle to incoming data. He is further committed, it is alleged, to the equally unrealistic supposition that all those items of data must be absolutely certain, since no allowance is made for conditioning on data which do not have this apodeictic character. We shall show that this multiple objection is a consequence of multiple confusions about the claims and aims of the subjective Bayesian approach, and that when these are dissipated, so too are the objections with them.

The final objection we shall discuss is that even were there nothing else wrong with this Bayesian theory, the empirically demonstrable fact is that people simply do not make their considered judgments according to its prescriptions. Were this true, and the evidence seems unequivocal, then it would appear to follow that the status of the subjective Bayesian theory as explanatory of the methodological evaluations people actually make must be severely in doubt. We shall show in due course that this conclusion is incorrect; let us now proceed to review all these objections in turn.

■ b THE BAYESIAN THEORY IS PREJUDICED IN FAVOUR OF WEAK HYPOTHESES

Discussing theories of inductive inference which assess the empirical support of hypotheses by changes in their probabilities on receipt of the relevant new data, Watkins (1987, p. 71) asserts that such theories are "prejudiced" in favour of logically weaker hypotheses. This is a favourite charge of the Popperian school and is frequently made by its eponymous founder; for example, Popper (1959, p. 363; his italics) writes that "[scientists] have to choose between high probability and high informative content, since *for logical reasons they cannot have both*".

Such a charge is quite baseless. There is *nothing* in logic or the probability calculus which precludes the assignment of even probability 1 to any statement, however strong, as long as it is not a contradiction. The only other way in which probabilities depend on logic is in their decreasing monotonically from entailed to entailing statements. But this again does not preclude anybody from assigning any consistent statement as large a probability as they wish. Popper's thesis that a necessary concomitant of logical strength is low probability is simply incorrect.

Glymour attempts to argue a variant of Popper's objection, but this too is easily repulsed. Glymour claims that since the observable consequences of scientific theories are at least as probable as the theories themselves, then in a Bayesian account one is unable to account for our entertaining theories at all:

> On the probabilist view, it seems, they are a gratuitous risk. The natural answer is that theories have some special function that their collection of observable consequences cannot serve; the function most frequently suggested is explanation.... [But] whatever explanatory power may be, we should certainly expect that goodness of explanation will go hand in hand with warrant for belief, yet if theories explain and their observational consequences do not, the Bayesian must deny the linkage. (Glymour, 1980, pp. 84–85)

The Bayesian certainly does want to justify the quest for theories in terms of a desire for explanation that a congeries of observational laws cannot by itself provide; but he would also, for very good reason, deny the linkage Glymour alleges between explanatory power and warrant for belief. Indeed, counterexamples to the claim that any such linkage exists are only too easy to find: a tautology, to take an obvious one, has maximal warrant for belief and minimal explanatory power. This does not, of course, imply that what we take to be good explanations do not tend to have correspondingly high probabilities on the available evidence. They do. But Glymour's premiss makes the additional claim that an increase in "warrant for belief" should imply an increase in explanatory power. That premiss is clearly false, and Glymour's objection collapses.

It is odd that Glymour and the Popperians should converge in charging Bayesians with an implicit denial of the value of deep explanatory theories but take as their points of departure opposed positions: Glymour thinks that good explanatory theories by that token justify a correspondingly large claim to belief, and the Popperians assert that such theories merit the lowest possible degree of belief. Whatever their starting points, however, the charge of Glymour, Popper, et al. that Bayesians must in principle undervalue theories is patently false. Perhaps a homely analogy will dispel any lingering doubts that may remain. A jury has always at least two mutually inconsistent hypotheses to consider: that the accused is guilty, and that the accused is not guilty and there is some alternative explanation of the known facts. They wish to determine which, relative to those facts, is the more probable hypothesis. Imagine their surprise at being informed that, since they wish to determine the more probable hypothesis relative to the available data, they are thereby committed on their return to the court to announcing that their favoured conclusion is the statement of the factual data they have been given (*see also* Horwich, 1982, p. 132). Scientists, like the court, want information of a specific sort combined with the assurance that it is credible information; and these demands *can*, despite Popper's solemn asseverations to the contrary, simultaneously be met, and the Bayesian theory tells us how.

■ c THE PRIOR PROBABILITY OF UNIVERSAL HYPOTHESES MUST BE ZERO

Popper, we noted in the previous section, asserts that it is impossible for a hypothesis to possess both high informative content and high probability. In particular, he asserts that the probability of a universal hypothesis must, for logical reasons, be zero (1959a, appendices *vii and *viii). He occasionally remarks (for example, 1959a, p. 381) that the constraints imposed by the probability calculus alone require that the only consistent assignment of a probability to such a hypothesis is zero.

Were Popper correct, then that would be the end of our enterprise in this book, which is to represent the procedures of inductive inference as consistency constraints on the assignments of subjective probabilities. For the truth of Popper's thesis would imply that we could never regard unrestricted universal laws as confirmed by observational or experimental data, since if $P(h) = 0$, then $P(h \mid e) = 0$ also, whatever finite sample data e may consist of. But Popper's thesis is untrue. Even in Carnap's so-called continuum of inductive methods (Carnap, 1952; *see* our discussion in Chapter 3), characterised by the values of a non-negative real parameter λ, one of those methods (corresponding to $\lambda = 0$), assigns, in an obviously consistent way, positive probabilities to a class of strictly universal hypotheses over an infinite domain; and, as we noted in Chapter 3, Hintikka's systems of inductive logic almost invariably assign positive prior probabilities to consistent universal sentences, whether the domain of individuals is finite or infinite.

Popper's arguments for his zero-probability claim are really designed to show something considerably less ambitious than the vastly overstrong thesis that there can be no consistent assignment of a nonzero probability to a universal hypothesis; what they aim at showing, as an examination of his text reveals, is that the so-called *logical* probability of a universal hypotheses must be zero (Popper, 1959, p. 364; a critical examination of these arguments is to be found in Howson, 1973 and 1987). We have already (Chapter 3) discussed the thesis that there is a genuine quantity, the logical probability of a sentence a, and concluded that the assumption that there is involves an unacceptably arbitrary, and hence most 'unlogical', degree of apriorism; and a uniform assignment of probability zero to non-tautological universal hypotheses is to our mind no less arbitrary than any other assignment.

Nevertheless, Popper has called our attention to a matter which deserves some comment, namely the fact of the apparent claim to complete generality of much of science and the episodic character of its history, successively punctuated by the demise of great explanatory theories. In view of these facts, we should, it seems, expect all current theory eventually to be overthrown by some new data, new candidates to emerge, become refuted in their turn, and so on ad infinitum. However, it is far from clear that such bleak pessimism really is the lesson taught by the history of science. The mere fact that succeeding extensions of the observational base of science have caused the demise of many an explanatory theory does not demonstrate the appropriateness of total scepticism, nor does it even make it plausible. If up till now I have failed to find the thimble, I do not conclude, and certainly ought not to conclude, that further quest is hopeless. Of course, science is not hunt the thimble, but this does not destroy the point of the analogy, which is that a number of past failures to discover the truth does not by itself imply that one will not one day be successful.

Pessimism on that particular score is certainly not something to which the practitioners of science themselves seem to subscribe. They are not discouraged by the record of others' failures: there is a great deal of biographical and anecdotal evidence which suggests that, on the contrary, some of the most illustrious of scientists not merely invest positive levels of confidence in their theories but, at any rate initially, are frequently prey to the wildest optimism. Watson and Crick quickly were totally convinced that they had discovered the structure of the DNA molecule, to take a well-documented example. This may not be global physics, but even there the picture is hardly one of unrestrained scepticism. Einstein's confidence in the correctness of his approach notoriously bordered on the hubristic. For example, when, after the reports of the 1919 eclipse expedition, someone asked Einstein what he would have done had the result not been confirmatory of his theory, he replied "Then I would have to pity the dear Lord. The theory is still correct". Even where scepticism supervenes, but accompanied by a determination to persist with the theory—the continuing and fruitful applications of non-relativistic quantum mechanics are a case in point—any subsequent predictive success will usually be attributed to some characteristic feature of that theory even if the particular formulation is thought to be false. But this simply means that it is a suitably modified form of that theory which is regarded as being confirmed. To sum up, there is no evidence that people regard general theories as invariably false, and no evidence that they ought to.

Before we turn to new matters, we must consider the objection that we ourselves are forced, by the way in which we choose to understand Bayesian probabilities, to assign probabilities of zero to universal hypotheses. For we have interpreted these probabilities as implicit assertions about fair betting quotients; to be precise, we regard the assignment of a probability p to h to mean that the individual making that assignment regards p, in the light of his background information, as a fair betting quotient. But if h is universal, then any bet on h might be lost but could not ever be won; consequently, or so it seems, the only fair betting quotient on h is zero, for only the value zero confers no advantage to either side of the bet.

We ourselves in Chapter 3 brought this objection, it may be recalled, against the standard employment of the Dutch Book Argument to show that degrees of belief ought to satisfy the probability calculus. But our use of that same argument is not open to the objection. For we have defined X's degree of belief in the truth of h as what X thinks is the fair betting quotient on h, on the (possibly counterfactual) assumption that the truth-value of h were to be unambiguously decided. That in this sort of case that assumption *is* counterfactual is irrelevant. This is not merely an ad hoc method of evading the problem either, as it might at first sight seem. In fact, it only seems so because of the legacy of Ramsey, Savage, and others, who insisted on behavioural criteria for determining degrees of belief. From the point of view of someone attempting to elicit his or her own strength of belief, on the other hand, introspective thought-experiments, like the one we ourselves invoked, are quite unexceptionable.

■ d PROBABILISTIC INDUCTION IS IMPOSSIBLE

This dramatic claim is made by Popper and David Miller (1983), who also supply what purports to be a rigorous proof of it. Their argument is as follows. According to Bayesian theories of support or confirmation, whether they are subjectively based or not, evidence e supports hypothesis h if and only if $P(h \mid e) > P(h)$. Suppose that h entails e, modulo background information including initial conditions and so forth. Then it is easy to see that e supports h if and only if $P(h) > 0$ and $P(e) < 1$. Suppose that these latter conditions are satisfied also, so that h is (it seems) supported by e. Popper and Miller demonstrate that if in addition $P(h) < P(e)$, then $\sim e \lor h$ is *counter-supported* by e, in the sense that its posterior probability relative to e is *less* than its prior probability (the proof of their result is very straightforward and we shall leave it to the reader to check if they so wish).

This simple theorem of the probability calculus is given a dramatic significance by Popper and Miller. For they claim that $\sim e \lor h$ represents the excess content of h over e, and interpret their result as stating that that excess content is always counter-supported by e. But e may well support h itself; as we saw, it does if h entails e. The Bayesian finds nothing in itself troubling in this breach of what Hempel called the Consequence Condition (that if e confirms h it confirms every consequence of h); he just thinks that the Consequence Condition is false. What does, or ought to trouble him, however, is the explanation of this breach which Popper and Miller provide. For on their interpretation of their formal result, the support e appears to give h is really just the self-support e gives e which is, after all, a part of the content of h. Indeed, if we measure the support $S(h,e)$ of h by e by the simple difference $P(h \mid e) - P(h)$, then it is not difficult to show that $S(h,e) = S(\sim e \lor h,e) + S(e,e)$ when h entails e. All support, conclude Popper and Miller, is really, therefore, self-, or what they call *deductive*, support (since e

entails e); the genuinely inductive component $S(\sim e \vee h, e)$ is always negative. So when we think evidence supports a hypothesis, we are, according to Popper and Miller, being misled; it really only supports that part of the hypothesis actually entailed by the evidence, the remainder of the hypothesis being counter-supported.

But Popper's and Miller's argument depends crucially upon identifying the excess content of h relative to e as $\sim e \vee h$, and as Redhead (1985) points out, there is an excellent and simple reason for not doing so. This is that $\sim e \vee h$ is actually a very weak statement (it is entailed by $\sim e$, for example), and h certainly has consequences which are consequences neither of e nor of $\sim e \vee h$: h itself is one of them. This simple fact in our opinion completely demolishes Popper's and Miller's premise that $\sim e \vee h$ "contains everything in h which goes beyond e" (1983, p. 687), and consequently leaves their anti-inductive conclusion quite unsupported.

It is not an adequate rejoinder (nor is it one which Popper and Miller make, incidentally), that since $S(h,e)$ can be split into two additive factors $S(\sim e \vee h, e)$ and $S(e,e)$, this by itself shows that e and $\sim e \vee h$ exhaust the content of h. All that the decomposition shows is that the values of the function $S(h,e)$ can be represented as the sum of the values of the functions $S(\sim e \vee h)$ and $S(e,e)$. The existence of such decomposition certainly does not tell us that the content of h is decomposed into $h \vee \sim e$ and e, since $S(h,e)$ is also decomposable into the sum of the functions $S(h \, \& \, e, e)$ and $S(h \, \& \, \sim e, e)$ (Dunn and Hellman, 1986). Moreover, neither $h \, \& \, e$ nor $h \, \& \, \sim e$ are in general consequences of h, and hence cannot plausibly be regarded as being in the content of h (Dunn and Hellman disagree, incidentally; but see Howson, 1989).

Popper's and Miller's reason for identifying $\sim e \vee h$ as the excess content of h over e is that the conjunction of e and $\sim e \vee h$ is equivalent to h (when h entails e), and that $\sim e \vee h$ and e share no non-tautologous consequences. They argue (1987) that any consequence k of h which is not a consequence either of e or of $\sim e \vee h$ shares non-tautologous consequences with e ($e \vee k$ is one). But this last fact does *not* entail that k is not in the excess content of h over e; it is certainly in the set-theoretic difference of the consequence classes of h and e, and Popper and Miller themselves define the content of a statement to be its set of consequences. The fact that k has consequences in common with e is irrelevant.

But Popper and Miller also employ a numerical measure *ct* of content, *ct(h)* being defined as *P(~h)*, where *P* is what they call a "logical" probability function. And this measure, which is widely endorsed, has the property, as the reader can easily check, that *ct(e)* + *ct(~e* ∨ *h)* = *ct(h)*, where *h* ⊢ *e*, which would seem to reinforce their claim that *e* and *~e* ∨ *h* do exhaust the content of *h*. Hintikka, who discusses the properties of this measure, anticipates Popper and Miller in concluding that *~e* ∨ *h* is correctly identified as the excess content of *h* over *e* (1968, p. 313). But we should not be at all impressed by this additivity property of *ct*: all it really means is that *ct* is too coarse-grained to 'notice' that consequences additional to those of *e* and *~e* ∨ *h* separately are created on conjoining *e* with *~e* ∨ *h*, a fact which is symptomatic of its quite general inability to register the additional content created by the conjunction of two logically independent statements.

Let us briefly justify this last remark. It is easy to show that

$$ct(a) + ct(b) - ct(a \lor b) = ct(a \,\&\, b).$$

Suppose that *m* is a measure on the set of sets of sentences of the language from which *a* and *b* are drawn, and suppose that we identify *ct(c)* with *mCn(c)*, where *Cn(c)* is the set of consequences of the sentence *c*: in other words, suppose that we regard *ct* as measuring consequence classes. *Cn(a* ∨ *b)* is equal to the intersection of *Cn(a)* and *Cn(b)*, and it follows that *ct* assigns zero measure to the net increase of consequences created by conjoining *a* and *b*. This is best seen graphically in the diagram below, where the baseless cones whose vertices are *a* and *b* represent the consequence classes of *a* and *b* (the shaded area represents *ct(a* & *b)*).

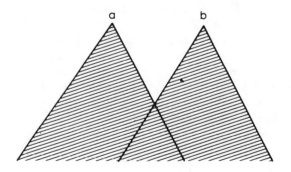

If one thinks that numerical measures of content are worth investigating at all, then, in view of these observations, one ought to conclude that *ct* is a very *in*adequate and misleading measure of content—though in view of our earlier doubts about whether any useful notion is referred to by the term 'logical probability', any person-independent notion of content based on a probability function seems to be ruled out. Howson and Franklin (1986) investigate numerical content-measures which reflect more faithfully the structure of the underlying consequence classes than does *ct* and show that in this respect more adequate measures are afforded by either $\text{Inf}(h) = -\log P(h)$, the so-called information measure (its expected value is Shannon's entropy), or $\dfrac{1 - P(h)}{P(h)}$, or the odds against h based on P. Whatever the eventual value of evolving measures of content based on 'suitable' probability functions, we can conclude that the properties of *ct* provide no support for Popper's and Miller's anti-inductivist strictures; nor, as far as we can see, is any to be derived from any other quarter.

■ e THE PRINCIPAL PRINCIPLE IS INCONSISTENT

David Miller (1966) produces an ingenious argument which appears to demonstrate that the principle which, following Lewis, we have called the Principal Principle, is inconsistent. According to that principle, on which the Bayesian analysis of statistical inference rests, the (subjective) probability that an event described by the sentence a will occur at a particular trial of type T is equal to r, if our data are confined to the information that the physical probability of a, relative to the conditions T, is r. We can write this concisely as the equation

(1) $P(a_t \mid P^*(a) = r) = r$

where P is the degree-of-belief probability and P^* the physical probability, and a_t is the statement that a, a generic event-description like 'the coin lands heads 3 times in a sequence of 5 consecutive tosses', is satisfied by the outcome of the particular trial taking place at time t.

Miller's argument is as follows. Let r be $\frac{1}{2}$. Then by (1)

$$P\left(a_t \mid P^*(a) = \frac{1}{2}\right) = \frac{1}{2}.$$

But clearly, $P^*(a) = \frac{1}{2}$ if and only if $P^*(a) = P^*(\sim a)$; and the probability calculus tells us that we can substitute equivalent statements, whence we obtain

(2) $P(a_t \mid P^*(a) = P^*(\sim a)) = \frac{1}{2}$.

However, we can also instantiate (1) thus:

(3) $P(a_t \mid P^*(a) = P^*(\sim a)) = P^*(\sim a)$

and combining (2) and (3) we infer that $P^*(\sim a) = \frac{1}{2}$, which is odd since no factual premise of any kind has been employed in the derivation. While this result may not have the form of an outright contradiction, it very quickly leads to one. For we can repeat the reasoning above with the two substitution instances $P^*(a) = \frac{2}{3}$ and $P^*(a) = 2P^*(\sim a)$ instead of $P^*(a) = \frac{1}{2}$ and $P^*(a) = P^*(\sim a)$ respectively, whence we infer that $P^*(a) = \frac{2}{3}$; and this is in explicit contradiction to $P^*(a) = \frac{1}{2}$.

Were Miller's derivation formally sound, the consequences for the Bayesian theory of statistical inference would be little short of disastrous; for the Principal Principle provides, as we saw in Chapter 9, the means of evaluating the likelihood terms $P(e \mid h)$ in Bayes's Theorem. Without the principle, Bayes's Theorem would merely contain three undetermined terms, $P(h)$, $P(e)$, and $P(e \mid h)$, where these are either probabilities or probability densities, and would yield no information at all about the value of the posterior probability or probability-density $P(h \mid e)$. The characteristic and often striking properties of the posterior probabilities of statistical hypotheses are due to the behaviour of the likelihood function $g(i,n) = P(e_n \mid h_i)$, where n is the sample size, and $(h_i)_i$ is a family of alternative hypotheses about the value of the physical probability distribution, indexed by a parameter i characterising this distribution, over a sample space one of whose outcomes is e_n.

But, as we also noted in Chapter 9, odds different from those based on the Principal Principle are demonstrably unfair, and this tells us that *something* must be wrong with Miller's clever inconsistency proof. The question is, what? It is certainly not very easy to spot his error, and a considerable number of eminent people have disagreed amongst themselves as to where the error lies. Jeffrey (1970) lists, accompanied by his own, the contemporary analyses of the paradox, though what we believe to be the correct solution was not found until 1979. It is expounded in Howson and Oddie (1979), and we shall reproduce it briefly now.

Miller's error is difficult to spot because it is concealed by the notation which, precisely in consequence of its being well adapted to smooth exposition and development of the theory, does not make explicit all the distinctions which are nevertheless implicit. The erroneous step in Miller's derivation is taking (3) to be a substitution instance of (1). (3) is *not* a substitution instance of (1); it makes a quite different *type* of assertion from (1), and it will help the reader see why if they first turn back to and re-read Chapter 2, section **f.1,** where random-variable-statements are introduced and their meaning discussed. For (1), though it may not look like it, is an equation involving random variables.

This fact is obscured by our tendency to regard $P^*(a)$ as a number. But in the context of a discussion in which $'P^*(a) = r'$ is itself a statement assigned a probability value (by the function P) $P^*(a)$ is not a number: it is something which takes a range of possible values—those possible values being, of course, all the real numbers in the closed unit interval. And a quantity which takes different values in different possible states of the world, and over whose values there is a probability distribution, is a random variable. Let us accordingly replace the term $P^*(a)$ in (1) by X, so that (1) becomes

(4) $P(a_t \,|\, X = r) = r,$

for all r, $0 \le r \le 1$. Now recall, from Chapter 2, that we can replace the sentences in (4) by the sets of possible worlds making them true; no formal difference exists between the linguistic and the set-theoretic representation of a class of sentences. So we can rewrite (4) as

(5) $P(M(a_t) \,|\, M(X = r)) = r.$

Let us now concentrate on $M(X = r)$. Looking back at Chapter 2 again, we see that $M(X = r) = \{w : X(w) = r\}$, where the w's are the members of the outcome space of the stochastic experiment relative to which the distribution P^* is defined. So (5) can be written

(6) $P(M(a_t) \,|\, \{w : X(w) = r\}) = r.$

But (3) has the form

(7) $P(M(a_t) \,|\, \{w : X(w) = Y(w)\}) = Y(w)$

where Y is another random variable equal to $1 - X$; Y is of course $P^*(\sim a)$ and $P^*(\sim a) = 1 - P^*(a)$. But it is now obvious that (7) is not a legitimate substitution instance of (5); it contravenes the logical rule that terms involving so-called free variables, like w, must not be substituted into contexts in which those free variable become bound, as the operator $\{w:\ldots\}$ binds $Y(w)$. (*See* Mendelson, 1964, p. 48, for the statement of this rule for logical quantifier operators.) It follows that (3) is not a legitimate substitution instance of (1) and the derivation of Miller's paradoxical conclusion cannot proceed. The Principal Principle is consistent.

■ f HYPOTHESES CANNOT BE SUPPORTED BY EVIDENCE ALREADY KNOWN

This objection is due originally to Clark Glymour (1980), and has since been echoed by many others. Let us quote from Glymour.

> Newton argued for universal gravitation using Kepler's second and third laws, established before the *Principia* was published. The argument that Einstein gave in 1915 for his gravitational field equations was that they explained the anomalous advance of the perihelion of Mercury, established more than half a century earlier.... Old evidence can in fact confirm new theory, but according to Bayesian kinematics it cannot. (p. 86)

By "Bayesian kinematics" Glymour here means simply the principle that one's degree of belief in a hypothesis h in the light of evidence e is equal to one's conditional degree of belief in h relative to e, which, if one is coherent, is a conditional probability $P(h \mid e)$. Glymour's thesis that Bayesian kinematics cannot account for the confirmation of new theories by old facts is grounded on the undoubted fact that relative to a stock of background information including e, $P(e)$ is 1, whence $P(e \mid h)$ is 1 also, so that it follows immediately from Bayes's Theorem that $P(h \mid e) = P(h)$. Thus e does not raise the prior probability of h and hence, according to the Bayesian, does not confirm it.

Though Glymour's reasoning appears to be sound, it has the rather strange consequence that *no* data, whether obtained before or after the hypothesis is proposed, can, within a Bayesian theory of confirmation, confirm *any* hypothesis. For even if the hypothesis h is proposed before evidence e is collected, then by the time someone comes to do the Bayes's Theorem calculation, the terms $P(e)$, $P(e \mid h)$ must again be set equal to

1, since by that time *e* will of course be known and hence in the contemporary stock of background information. But it would be absurd to infer that according to the Bayesian theory *e* did not support *h*, and not even the most committed opponent of that theory would claim the theory damaged by this demonstration, for it is clear that the theory has been incorrectly used and that the mistake lies in relativising all the probabilities to the *totality* of current knowledge: they should have been relativised to current knowledge minus *e*. The reason for the restriction is, of course, that *your current assessment of the support of* h *by* e *measures the extent to which the addition of* e, *to the remainder of what you currently take for granted, would cause a change in your degree of belief in* h.

In other words, once *e* has become known and you want to assess the support *e* gives *h*, the probabilities $P(e \mid h), P(h)$, and $P(e)$ which you will consequently want to evaluate are relativised to the counter-factual knowledge-state in which you still do not know *e*. Once this is grasped, the solution to the problem of how to deal with the new hypothesis/old evidence problem is obvious in principle, though in some cases difficult to apply in practice: you relativise the probabilities, so far as you are able, to what you currently know minus the data whose current confirmatory capacity vis-à-vis the hypothesis you want to establish.

Glymour considers this reply quite sympathetically, but contends that nevertheless in

> actual historical cases...there is no single counterfactual degree of belief in the evidence ready to hand, for belief in the evidence sentence may have grown gradually—in some cases it may have even waxed, waned, and waxed again. (p. 88)

He cites as an example the data on the perihelion of Mercury; there were different values obtained for this, over a period of several decades, by different methods, and employing mathematical techniques sometimes without rigorous justification. Glymour contrasts this situation with the results of tossing a coin a specified number of times, where he thinks it does make sense to talk of the probability of that outcome, as if it had not yet occurred. But in the case of Mercury's estimated perihelion advance, "there is no single event, like the coin-flipping, that makes the perihelion anomaly virtually certain" (Glymour, 1980, p. 88).

But whether there is as much epistemic warrant for the data in 1915 about the magnitude of Mercury's perihelion advance as there is about the number of heads we have just observed in a sample of a hundred tosses of a coin is beside the point. About some data we may be more tentative, about other data less. The Bayesian theory we are proposing is a theory of inference from data; we say nothing about whether it is correct to accept the data, or even whether your commitment to the data is absolute. It may not be, and you may be foolish to repose in it the confidence you actually do. The Bayesian theory of support is a theory of how the acceptance as true of some evidential statement affects your belief in some hypothesis. How you came to accept the truth of the evidence, and whether you are correct in accepting it as true, are matters which, from the point of view of the theory, are simply irrelevant. Glymour's disquisition on the frailty of much scientific data is therefore, however valuable in its own right, beside the point of evaluating the adequacy of the Bayesian theory of inference.

The same is true of his subsequent discussion of the lack of any *general* means of computing these degrees of belief. He considers and rejects several candidates, and concludes that it may not be some "old result that confirms a new theory, but rather the new discovery that the new theory entails (and thus explains) the old [result]" (Glymour, 1980, p. 92). This observation prompts Glymour to propose a new, quasi-Bayesian criterion for old evidence e to be taken as confirming a new theory h, namely that

(8) $P(h \mid e \ \& \ (h \vdash e)) > P(h \mid e)$

where the probability calculus is weakened appropriately, by replacing the conditions on axioms 2 and 3 by 'if it is known that t is a tautology then ...', and 'if it is known that $a \vdash \sim b$ then ...' (we have modified Glymour's notation in (8) slightly and omitted explicit reference to background information). This emendation of the classical theory is sympathetically endorsed by Niiniluoto (1983), and is further examined by Garber (1983), though the same idea seems first to have been proposed and developed by I. J. Good (starting with Good, 1948), who calls the resulting notion of probability "dynamic", or "evolving" probability.

Let us consider these issues in turn: first, the alleged absence of any general rule for computing $P(e)$ in the counterfactual way suggested. We can ignore this discussion because the sort of Bayesianism we are advocating is not regarded as a source of rules for computing all the probabilities in Bayes's Theorem. In particular, we are under no obligation to legislate concerning the methods people adopt for assigning prior probabilities. These are supposed merely to characterise their beliefs subject to the sole constraint of consistency with the probability calculus. In the context of this particular discussion, all we are concerned with is that people are capable in many cases of determining, possibly only very roughly, to what extent they think a piece of data likely relative to a stock of residual background information.

What of Glymour's suggestion that it is not simply the knowledge of e, but in addition the discovery that h explains e, on which one bases the conclusion that h is confirmed by e? This is undoubtedly true; the question, which Glymour thinks is to be answered in the affirmative, is whether this should prompt any change in the formalism of the Bayesian theory if the latter is to reflect it adequately. On the standard account, the knowledge that h entails e is discharged in computing $P(e \mid h)$ to be 1, and allowance for the recognition of entailment relations is made generally in this way. It is true, however, that this has often been deemed unrealistic, because it implies, to take a dramatic example, that the degree of belief of any coherent individual in the four-colour conjecture, conditional on the truth of certain other mathematical statements, is 1, and always was 1 from the time the conjecture was first formulated; whereas, of course, it was exceedingly difficult, and took a great deal of time (including computer time) to prove that result.

But we are not asserting that everybody ought to be consistent, nor that people actually are invariably consistent. The foundation of subjective Bayesianism is probabilistic consistency, admittedly, but this is an ideal which is not always attainable, and when complex logical entailments are involved, we should not expect that ideal to be realised in more than a small number of cases, if at all. It is certainly no part of our use of the theory as an explanatory tool that people are invariably consistent any more than it is the case that somebody who wished to provide an account of deductive inference is committed to the view that people either ought to be or are invariably deductively consistent. All we claim is that when people recognise or are apprised of deductive relationships between hypothesis and evidence, they often draw conclusions about levels of support in accordance with those determined within the Bayesian theory by the same initial probabilities.

This means that Glymour's non-classical condition (8) does not after all make for a markedly, if at all, more realistic theory of Bayesian inference. Nor is it necessary to introduce that confirmation condition (8), in the context of a weakened probability calculus, in order to give an adequate Bayesian account of how old evidence can confirm new theories. On this account, recall, the probabilities which are used to determine support are relativised not to one's total background information, but to what that state of background information would be were one not yet to know the data in question. While Glymour's objections to this proposal are, we have argued, unconvincing, there remain two further objections which we must now consider.

The first is that relativisation of all the probabilities to what is strictly a fictitious state of background information is simply ad hoc: it is a device which avoids the otherwise embarrassing necessity of setting $P(e)$ and $P(e \mid h)$ equal to 1, but it does so at the cost of being in conflict with core Bayesian principles. But a little reflection should convince the reader that this charge is untrue. Core Bayesian principles simply state the conditions—obedience to the probability calculus—for a set of degrees of belief, relative to a stock of background information, to determine a corresponding set of odds which are not demonstrably unfair. There is absolutely nothing in this which asserts that in computing levels of support, one's subjective probabilities must define degrees of belief relative to the totality of one's current knowledge. On the contrary; as we pointed out earlier, the support of h by e is gauged according to the effect which a knowledge of e *would* now have on one's degree of belief in h, on the (counter-factual) supposition that one does not yet know e.

The second objection occurs in a paper by Campbell and Vinci (1983). Their argument is in essentials as follows. Suppose first of all that h predicts, relative to suitable initial conditions, an event e, and the experiment designed to elicit e if h is true has not yet been performed. Suppose also that relative to current background information, $P(e)$ is high. The experiment is duly performed and e is observed. The support of h by *e* is inconsiderable, because of course *P(e)* is high. Now suppose that *P(e)* is high precisely because *e* describes the same sort of effect as has already been observed to occur in contexts which are thought to be strongly analogous to those in which *e*'s occurrence is predicted by *h*. Let the conjunction of all this past data be *e'*. Suppose that *h* was proposed well after *e'* was known, and that *h* also 'predicts' *e'* in the relevant circumstances, although it was not deliberately designed to do so (the reason for this rider will only become fully apparent when we come to section *g*). Finally, suppose that relative to background information minus *e*, *P(e')* is low. According to our analysis, the support of *h* by *e'* is considerable, unlike its support by *e*.

Campbell and Vinci flesh out the picture by taking *h* to be statistical thermodynamics and *e* and *e'* to describe examples of Brownian motion in different liquids. *Their* conclusion, for which incidentally they produce no argument but seem to take as intuitively justified, is that the support of *h* by *e* ought to be no different from its support by *e'*, which they agree is high (p. 323). But this conclusion is far from intuitively clear, and indeed seems to us to be, on the contrary, intuitively quite wrong: what we are seeing here is, after all, the phenomenon of diminishing support by repeated occurrences of the same type of effect. Initially *(e')* the support is high; later *(e)* it is not. This 'intuition' (for want of a better word) is reinforced by our analysis. There is not much more that we can say. We have given reasons, which seem to us good ones, for a conclusion disputed by Campbell and Vinci on grounds which presumably are supposed to be intuitive, but to us seem anything but.

■ g HYPOTHESES ARE NOT SUPPORTED BY DATA THEY WERE CONSTRUCTED TO EXPLAIN

There is one feature of the Campbell–Vinci example which may well cause the reader disquiet; indeed, it may appear that the discussion of that example implicitly refutes the Bayesian theory of support. The feature we are referring to is the innocent-looking clause stipulating that *h* was not designed to accommodate *e'*. The attentive reader may have noted that according to our analysis of the situation, *h* would be supported by *e'* to exactly the same extent *whether this condition were satisfied or not*. Nothing in the method of computing this can take account of whether h were deliberately constructed with the explanation of e' in mind or not.

We have already discussed in Chapter 4 the claim that a hypothesis designed to fit some piece of data is not supported by it to as great an extent as one which also fits the data but accidentally, so to speak. We argued there that this claim was false; and the Bayesian theory of support is certainly inconsistent with it. But there are arguments for the view, and these both sound convincing and also number among their subscribers many if not most contemporary philosophers of science. We shall examine these arguments now and show that their plausibility vanishes on closer inspection.

Suppose that data e are employed as a constraint in constructing a deterministic hypothesis h in such a way that h is made to entail e if suitable initial conditions are met. It seems fairly clear that h is not under risk of falsification by $e;$ and in consequence of this fact it is argued that e cannot support h. The argument is a popular one; thus Giere writes

> If the known facts were used in constructing the model and were thus built into the resulting hypothesis...then the fit between these facts and the hypothesis provides no evidence that the hypothesis is true [since] these facts had no chance of refuting the hypothesis. (1984, p. 161)

Glymour (1980, p. 41) voices a substantially identical opinion, as does Zahar, in a slightly more elaborate way (1983, p. 245). But the argument is quite fallacious. First, note that it is simply false that any *fact* has a "chance" of refuting anything. If e is a factual statement and h a hypothesis, then e either refutes h or it does not, and it does or does not whether h was designed to explain or embody e or not. The confusion here is a confusion of a random variable (in effect, the experimental set-up E) with one of its possible outcomes, e. What has the chance of refuting h, if at least one of its possible outcomes is inconsistent with h, is E, and only E. And when it is E rather than e which is recognised to have this chance, the argument collapses, for even if h was constructed from one particular outcome of E, it is in general logically possible that E could have produced an outcome e' which is inconsistent with h.

Suppose that h entails e modulo initial conditions which obtain when the experiment E is performed. It is interesting to note that Giere's corrected principle—that if E stands no chance of refuting h, then h cannot be supported by the outcome e of E either—follows directly from Bayes's Theorem itself if we are allowed to render "chance" as "probability", for to say that E has no chance of refuting h entails that $\sim e$ has no chance of occurring, which translates into the condition that $P(e) = 1$. Since $P(e \mid h) = 1$, it follows that $P(h \mid e) = P(e)$ and that the support for h from e is zero. But that $P(e) = 1$ certainly does not follow from the fact that h was constructed to explain e.

This conclusion, however, is implicitly contested by both Giere (1984) and Redhead (1986), for they argue that if h is deliberately constructed to explain e, then the probability of e given that h is false is equal to 1. But from $P(e \mid \sim h) = 1$ it easily follows, given that $P(e \mid h) = 1$, that $P(e) = 1$, since $P(e) = P(e \mid h)P(h) + P(e \mid \sim h)P(\sim h) = P(h) + P(\sim h) = 1$. How do these authors manage to conclude what seems, then, clearly false, that $P(e \mid \sim h) = 1$ when h is constructed to explain e? Let us take Giere first. He considers Mendel's simple one-factor, two-allele model of inheritance in his pea plants, which Giere contends was constructed to explain the two-to-one ratio of tall to dwarf plants in the second filial generation after crossing true-breeding tall and dwarf in the parental generation, and then mating the tall offspring. Giere asserts that "fitting this case was . . . a necessary requirement for any model to be seriously entertained. So there seems no way [in which the data could be improbable given the negation of Mendel's hypothesis]" (1983, p. 118).

Giere provides no argument for this claim, however: on the contrary, he appears to take it as self-evident. But it is far from self-evident that Mendel's data would not be improbable were his own explanation of them to be false; indeed, as that was the *only* explanation which seemed plausible to Mendel, its falsity would presumably render those data, were they assumed to be still conjectural, relatively improbable as far as he was concerned. And this, as we have seen, is sufficient for a Bayesian to be able to explain the undoubted fact that Mendel himself took his data to be strongly confirmatory of his model. Giere has not justified his thesis—nor indeed could he—that $P(e \mid \sim h) = 1$ when h has been designed to explain e.

Let us now look at Redhead's argument for that claim. The curious thing about the argument he offers is that it purports to be a Bayesian argument; but the Bayesian position seems quite opposed to any such conclusion, as we have seen. But let us see what Redhead says. He commences by casting Giere's premise, that in seeking to explain e one is making the explanation of e a necessary condition for any model to be entertained, into the more explicitly Bayesian condition that the desire to explain e acts as a "filter" upon the set of all hypotheses, allowing a non-zero prior probability only to those which, relative to a suitable set of auxiliary hypotheses and initial conditions, call them a, entail e. It certainly follows from this "filter" condition that $P(e \mid h) = 1$, since

$$P(e \mid \sim h) = \frac{\Sigma P(e \mid h_i)P(h_i)}{P(\sim h)},$$

where the sum is over all the mutually exclusive hypotheses h_i whose disjunction is equivalent, with probability 1, to $\sim h$ (the disjunction may include what Shimony (1970) calls a "catch all" hypothesis, simply equivalent to the negation of the disjunction of all the remaining h_i and h). Redhead's filter condition ensures that all the h_i in the sum are such that $P(e \mid h_i) = 1$, and hence that $P(e \mid \sim h) = \frac{\Sigma P(h_i)}{\Sigma P(h_i)} = 1$. But this, of course, implies Giere's conclusion, that h receives null support from e, since $P(e \mid \sim h) = P(e \mid h) = 1$.

One obvious fault of Redhead's assumption that, in setting out to construct an explanation of e one is, in effect, assigning a positive prior probability only to those hypotheses which entail e relative to a, is that it is inconsistent. For a tautology does not explain e, yet its probability must be one. Even if the filtering is restricted to some partition $\{h_i\}$ (and it is not clear that anyone determines such a partition is attempting to explain e), the condition that only those members have positive probability which entail e is still far too strong. For it yields exactly the same, null-support, result for hypotheses proposed independently of the data e as it does for those deliberately designed to explain e. Suppose h is designed to explain e, modulo a. It follows from the filter condition that $P(e)$ equals 1, since $P(e) = P(e \mid h)P(h) + P(e \mid \sim h)P(\sim h) = P(h) + P(\sim h) = 1$. But since $P(e) = 1$ it also must be the case that $P(e \mid \sim h') = 1$, where h' is any hypothesis which entails e modulo a. It follows that $P(h' \mid e) = P(h')$, so that h' is not supported by e either. The unwelcome strength of Redhead's filter condition stems from its implicitly making $P(e) = 1$. As we have already pointed out, however, $P(e)$ is not equal to unity just because e is a known effect we wish explained.

We conclude that attempts to show that data which hypotheses have been deliberately designed to entail, as opposed to independently predicting, do not support those hypotheses fail. On the contrary, the condition for support, that $\dfrac{P(e \mid \sim h)}{P(e \mid h)}$ be small, may be perfectly well satisfied in many such cases. We have already (Chapter 4) looked at some simple examples, though no harm will come from repeating them here. (i) a box is known to contain a number k of red and blue balls. We want to know the proportion of red balls. We take the balls out one by one and note their colour, until the box empty. We note that there are r red balls. We formulate the hypothesis h that the proportion of red balls originally in the urn is $\dfrac{r}{k}$. h is constructed from the data, and nobody would presumably deny that it is supported by the data since, together with some uncontentious background information, the data entail h and vice versa; thus $\dfrac{P(e \mid \sim h)}{P(e \mid h)}$ is zero.

If this example is thought too trivial and 'unscientific', consider (ii): we have the same box, but use a different experiment to discover the proportion of red balls. We take from the box a sequence of n balls, noting the colour of each, replacing it and shaking the box. Suppose that s of the balls we observe in this way are red. We formulate the hypothesis h' that the proportion of red balls in the box is $\dfrac{s}{n} \pm \epsilon$, where ϵ is a suitable real number. h' too would be regarded by most people as a hypothesis well supported by the data. Here, however, $P(e \mid h')$ is not only not equal to 1: it will be very small indeed if n is greater than 20, say, because it will be obtained by applying the background model of a binomial distribution with probability parameter p in the interval $\dfrac{s}{n} \pm \epsilon$. But while $P(e \mid h')$ is small, $P(e \mid \sim h')$ is in general very much smaller. (If we take the binomial model as *the* model of the experiment, with a uniform prior probability distribution over the values of the binomial parameter, then $P(e \mid \sim h')$ can straightforwardly be computed using the theorem of total probabilities; even if the prior distribution is not uniform, then the same method will give a good approximation for sufficiently large n.) The reference to balls in a box can be dispensed with in evaluating the significance of the example in (ii), incidentally, which is modelled in a great variety of estimably scientific experiments.

It might be objected that in regarding the hypotheses in (i) and (ii) as well supported by the data from which they were calculated, we are relying on background information containing well-defined models of the experiment: the data, in fact, merely perform the function of specifying parameters in those underlying models which are themselves taken to be independently very well-supported, to the point of their truth being taken for granted. This is true, but not damaging. In evaluating support we always and necessarily employ *some* background information which we take pretty much or even completely for granted. And this background information will always be analogous to a model with undetermined parameters, in that it will leave open a (more-or-less indefinite) range of alternative hypotheses about the structure of some experimental process. Of course, the data obtained from that experimental source will by no means uniquely determine which hypothesis is correct, given that background information; but then the data in (ii) above failed to fix the parameter uniquely either. Of course, also, cases of parameter fixing are very special examples of data-determining hypotheses: the space of possibilities is clearly defined, for one thing, with usually a simple mathematical structure (it is often an interval of real numbers). But these features do not at all affect the validity of our general conclusion, which is that in appropriate circumstances some data might both act as a constraint on the construction of hypotheses and simultaneously support those hypotheses; and the appropriateness of the circumstances can be characterised, albeit in general terms, as a function of the magnitude of the likelihood ratio $\dfrac{P(e \mid \sim h)}{P(e \mid h)}$.

■ h PREDICTION OR ACCOMMODATION?

There is a fairly ancient, many-sided debate about confirmation, one side of which asserts that hypotheses constructed deliberately to accommodate data e are never supported to the same (positive) extent by e as hypotheses which independently predict e. An extreme version of this view is that such data never support the corresponding hypotheses. The burden of the previous few paragraphs is that this extreme thesis is untenable. We have also argued (Chapter 4, section **j**) that the less extreme view (that had h independently predicted e, then it would have obtained more support from e than had it been constructed with e as an explicit constraint) is also false in general. However, we shall now exhibit some, not altogether atypical, circumstances in which the accommodating hypothesis receives at most as much support as the independently predicting one.

Suppose, for example, that h possesses an undetermined parameter a. An experiment is conducted to determine the value of a relative to the assumption that h is true. Its outcome is e. Let the resulting hypothesis, namely h together with the computed (or estimated) value of a, be h', and suppose that h', conjoined with the initial conditions of the experiment, entails e (it does not have to, in general). Now suppose that another hypothesis h'' had been formulated and proposed before that experiment had been performed, but together with the same initial conditions, it too entails e. Here we have two hypotheses, in other words, one (h') which has been made to accommodate e, and the other (h'') which independently predicts e. (It may, of course, be the case that fixing the parameter in h at some particular value enables h', the adjusted hypothesis, to explain e, but at the cost of making other predictions which are known to be false. The de Sitter modification of Poincaré's Lorentz-invariant gravitational theory is a case in point, where the parameter adjusted to yield the correct value of Mercury's annual precession caused the theory to be incorrect about the bending of light, among other things.)

Finally, suppose that prior to that experiment the prior probability of h'' is at least as great as that of h. Since we are now talking about relative magnitudes of support, we want some way of measuring supports on some numerical scale. We have already taken the function $S(\cdot,e) = P(\cdot \mid e) - P(\cdot)$, where the relevant hypothesis replaces the dot, as a natural measure which locates degree of support within the interval $[-1,1]$, and we shall use S in what follows. Where the initial conditions of the experiment are regarded as being part of the general background information, we have that $S(h',e) = \dfrac{P(h')[1 - P(e)]}{P(e)}$

and similarly that $S(h'',e) = \dfrac{P(h'')[1 - P(e)]}{P(e)}$. So the ratio of the supports of h' and h″ is just the ratio of their prior probabilities. However, h′ logically entails h, so that $P(h') \leq P(h)$, and by assumption $P(h) \leq P(h'')$. Hence $S(h',e) \leq S(h'',e)$.

We said that the condition for that inequality, namely that $P(h) \leq P(h'')$, is not too atypical. It often happens, for example, that one scientific theory predicts an effect e which serves only to fix a parameter-value in a rival theory. We are tempted to say that the first theory gets more support from e than does the second, even if the latter also 'predicts' e in those circumstances once the parameter has been fixed. But it is important to be clear about *why* we make this judgment. It is not because independent prediction *always* confers more support than accommodation, as an influential tradition commencing with Leibniz and including Whewell has claimed. Support depends on prior probability, as the support-function S makes clear, and a completely incredible theory will not in general be regarded as being supported whatever it predicts, or how it did so (in the limit we have a contradiction which predicts everything and is supported by nothing, for example). We make the judgment that the independently predicting theory h″ usually gets more support than the adjusted hypothesis h' precisely because h and h″ are rivals, and hence can be presumed to have comparable prior probabilities. And if those probabilities are equal, within the limits of imprecision which usually attends such judgments, then the inequality above will be valid.

These conclusions depend on the fact that the prior probability of h, the hypothesis with free parameters, determines the maximum support which is gained by h' from the data which fixed the values of the parameters: as we saw, $S(h',e) \leq \dfrac{P(h)[1 - P(e)]}{P(e)}$. This is a result of considerable significance. For it tells us, among other things, that a consequence of regarding the introduction of those parameters as *merely* an ad hoc way of accommodating the data, then the support of the resulting determinate hypothesis will certainly be inconsiderable. The plausibility of the thesis that predictions always glean more support than accommodations rests, we suspect, on nothing more than invalidly generalising from this special case in which the thesis is true (this point is made forcefully by Nickles, 1985, p. 200).

We can illustrate these remarks with an example we used earlier, in Chapter 4, section **k,** that of the new 'law' of free fall

(**) $s = g(t) + f(t)(t - t_1)(t - t_1) \ldots (t - t_n),$

where $g(t)$ is Galileo's law and $t_1, \ldots, t_n$ are the time instants at which observations were made of the corresponding values of s, and $f(t)$ is some arbitrary function of t which, we shall assume, is nowhere zero.

Clearly, (**) is obtained from the hypothesis that there are nonzero values of $a_1, \ldots, a_n$ such that

(***) $s = g(t) + f(t)(t - a_1)(t - a_2) \ldots (t - a_n)$

and the n observations; the latter uniquely determine the values $a_1 = t_1, \ldots, a_n = t_n$ of the parameters a_i once the functional form (***) is given. Intuitively (let us suppose that we are in the late sixteenth century), the prior probability of the parametric model (***) is zero or at most negligible in comparison with that of the Galilean law $g(t)$, because the parameters a_i and the function $f(t)$ are introduced purely ad hoc, for no reason whatever except that it is known that they will fit the observations.

(**) and (***) and the Galilean law itself are therefore instances of the hypotheses h', h, and h'' in our discussion above. We infer that the support of (**) by the n observations is smaller than that of the Galilean law, simply as a result of the distribution of prior probabilities over the parametric model (***) and Galileo's law. (**) is an extreme case of curve-fitting by adjustment of parameters whose introduction lacks any theoretical justification whatever.

To sum up: it is not true that data which are used as explicit constraints on the construction of explanatory hypotheses are thereby precluded from being counted in support of the resulting hypotheses. Nor is it true that a hypothesis which independently predicts the data gets more support from it than one which is constructed in order to predict it. The widespread belief to the contrary is an illicit extrapolation from those cases where the data-generated hypothesis is based on a theoretical model whose sole raison d'être is that it accommodates the observations, usually by being endowed with an appropriately large number of adjustable parameters. In these cases, where the theoretical model has no independent justification, support is withheld. The Bayesian theory here as elsewhere only articulates the feelings of the practitioners themselves: the economist Edgeworth, for example, expressed his reserve about Karl Pearson's fitting his (Pearson's) family of probability-density curves to the data, in the pointed question (Edgeworth 1895, p. 511) "what weight should be attached to this correspondence by one who does not perceive any theoretical reason for those formulas". Kepler fitted ellipses to Tycho's data for planetary orbits, but only after he had found independent reasons for that type of orbit. Nearer to home we find Kitcher castigating some sociobiologists' parameter adjustment on the ground that "the model gives absolutely no insight into the reasons behind the periodicity [the adjusted parameter] . . . the choice of a periodic function for the probability bears no relation to any psychological mechanisms" (1985, p. 375). And so on.

■ I THE PRINCIPLE OF CONDITIONALISATION, AND BAYESIAN LEARNING

Somebody who has degrees of belief $P(h)$, $P(e)$, and $P(h \& e)$ in the truth of the sentences h, e, and $h \& e$ thereby has, on pain of inconsistency, as we saw in Chapter 3, a degree of belief $P(h \mid e)$ in h, conditional on e's being true, where $P(h \mid e) = \dfrac{P(h \& e)}{P(e)}$. If e does turn out to be true, then the degree of belief of this person, again on pain of inconsistency, in h unconditionally becomes $P'(h) = P(h \mid e)$. This is the Principle of Conditionalisation whose validity we also proved in Chapter 3. It has, however, become a focus of critical attention in the past few years, and its status disputed. Hacking, Kyburg, Levi, and many others claim that the principle requires a justification independent of that of the axioms of the probability calculus, as we noted in Chapter 3; but we also observed there that the claim is false. However, there are some apparent objections to the principle, and these we must examine now.

One objection is that the Principle of Conditionalisation is the only mechanism in the Bayesian theory for learning by experience. Since the prior probability distributions which enter into the Bayes's Theorem expression of $P(h \mid e)$ themselves reflect what van Fraassen (1980) calls "the deliverances of experience", it would seem that they can achieve a Bayesian explanation only if they themselves are posterior probabilities, relative to some anterior reception of data and some yet prior probabilities; and so on, until an ultimately prior distribution is reached, prior to all empirical experience, far back in the history of the organism, at the dawn of its cognitive life. To explain how current empirical data affect current belief then seems to entail not only the reconstruction of the agent's successive acts of conditionalisation—a daunting if not practically impossible task—but also the characterisation of a state of primal ignorance; and we have seen in Chapter 3 that attempting to characterise primal ignorance in terms of some absolutely neutral prior distribution is a pretty hopeless task.

But none of this need worry us. There is nothing in the account we have given which commits us to the thesis that all change of belief takes place via Bayesian conditionalisation. In fact, that account implicitly contradicts such a thesis, for apart from anything else the data e appearing in the conditional probabilities are given exogenously: e is, for want of a better word, simply 'known'. The reader, aware of the fallibility of practically all 'deliverances of experience', may regard this admission as entailing a no less damagingly unrealistic account of inductive inference. We shall argue shortly that this is not so. But first things first. We are proposing a theory of inference; in particular, a theory in which from two inputs, e and a belief distribution, an output belief-distribution is generated by the Principle of Conditionalisation. Since we are not claiming that *all* belief distributions are obtained by conditionalisation, we are not committed to explaining the provenance of the input belief-distribution.

A more serious objection is that the data input e is simply taken as given. Many people have regarded this as an embarrassment for the Bayesian theory, because it has seemed to them that 'given' here is synonymous with 'certain'. Keynes distinguished between indirect, merely probable knowledge, arising from conditionalising on direct, or certain knowledge (1921, pp. 10–20), and in doing so set the terms of the subsequent debate to the extent that a quarter of a century later C. I. Lewis could remark that "If anything is probable, then something must be certain" (Lewis 1946, p. 186). Lewis's observation is, or so it would appear, fully endorsed by the Principle of Conditionalisation, since by that principle, $P'(e) = P(e \mid e) = 1$. However, as we conceded earlier, our knowledge of the world is not simply decomposable into two kinds, that which is infallible and that which is conjectural. If no one else, then Descartes should have made us aware that practically nothing is certain.

One well-known way of accommodating this objection, which yields the Principle of Conditionalisation as a special case, is due to Richard Jeffrey. Jeffrey's model for belief change as a result of experiential inputs ("probability kinematics") does not involve the ascription of any probability at all, let alone probability 1, to them. In the simplest possible case, we revise, as a result of some experience, our personal probability of one hypothesis h from $P(h)$ to $P'(h)$. How should this affect the probabilities of the various other hypotheses we contemplate? According to Jeffrey (1983, p. 169), if a is any other hypothesis then

(9) $P'(a) = P(a \mid h)P'(h) + P(a \mid {\sim}h)P'({\sim}h)$

Clearly, if $P'(h) = 1$, then we obtain the ordinary rule of conditionalisation, $P'(a) = P(a \mid h)$, so that "conditionalisation is a limiting case of the present more general method of assimilating uncertain evidence, and the case of conditionalisation is approximated more and more closely as the probability [of h] approaches 1" (p. 171). The case where more than one hypothesis has its probability exogenously altered is a straightforward generalisation of (9) (ibid.).

Both the Principle of Conditionalisation and Jeffrey's rule emerge as special cases in the theory of belief change presented in Williams (1980), in which the posterior distribution P of belief over a class of mutually exclusive hypotheses h_i is determined as that distribution which minimizes what Williams and others call the information in P relative to a prior distribution P^0, subject to whatever constraints are imposed as a consequence of some experiential input. Introduced by Hobson (1971) as the unique quantity satisfying some intuitively plausible desiderata for measures of relative information, the information in P relative to P^0, $I(P,P^0)$, is defined to be equal to

$$\sum_i P(h_i)\log\left(\frac{P(h_i)}{P^0(h_i)}\right).$$

I is intended to measure something like the probability-relevant magnitude of the information whose acquisition changes P^0 to P. Thus I is zero when P and P^0 are the same distribution (it is always nonnegative), and it becomes large without bound if P places an event close to 1 which P^0 places close to 0. It is also not difficult to see that the function P minimizing I, subject to the condition that for some statement a in the domain of P, $P(a) = 1$, is such that $P(h_i) = P^0(h_i \mid a)$ for every i. If instead we take the constraint, as in the Jeffrey situation, merely to be that $P(a)$ is some number between 1 and 0, then minimizing information subject to that constraint yields

$$P(h_i) = P^0(h_i \mid a)P(a) + P^0(h_i \mid \sim a)P(\sim a),$$

that is to say, we obtain Jeffrey's rule (Williams, 1980 p. 136).

A great deal has been written about the status of Jeffrey's rule. It should be emphasised that it is not the only way of generalising the Principle of Conditionalisation to contexts in which evidence is more or less uncertain: there is an infinity of ways. Nor is it justified by the sorts of consistency constraint that we have invoked to justify the probability axioms and the rule of conditionalisation (Armendt, 1980, provides however a Dutch Book argument for it, but only given certain other conditions). Nor, as Levi points out, is it clear how "to distinguish those initial shifts [from $P(h)$ to $P'(h)$] for which [the agent] has no justification from those [e.g. from $P(a)$ to $P'(a)$] which he can justify via an appeal to the initial shifts" (1967, p. 204). As far as Williams's rule is concerned, similar considerations apply. In particular, Williams offers no real argument why the appropriate posterior probability is the one which minimises I relative to the prior P^0. In summary, both Jeffrey's rule and Williams's generalisation are undoubtedly interesting and valuable developments in the Bayesian account, but as yet they remain speculative developments; their status as an extension of core principles needs more in the way of justification than they have yet got. For this reason nothing in our account will depend on their acceptance.

However, there remains the problem posed by the general fallibility of evidence statements and the fact that in the ordinary Bayesian account they are assigned posterior probability 1. There is a prima facie conflict here, as we noted above, which only some Jeffrey-type relaxation of the probability 1 condition can, it seems, avert. Despite appearances, however, there is really no conflict. In our account there is nothing that demands that what is taken as data in one inductive inference cannot be regarded as problematic in a later one. Assigning probability 1 to some data on one occasion does not mean that on all subsequent occasions it need be assigned probability 1. Levi (1967) has argued similarly, and persuasively, for just this point in his discussion of Jeffrey's rule. He also points out that it is a mistake to regard the ascription of probability 1 to a hypothesis as equivalent to the assertion that that statement is infallible. All that the ascription of probability one to *e* entails, in our and Levi's view, is that the agent takes *e* to be true in the light of his current experience. It does not follow that at some future occasion he might not have equally compelling reasons to regard *e* as false: *e* remains corrigible, in other words, but may quite reasonably, given appropriate background data, be currently assigned probability 1. Levi sums up the position very nicely:

> propositions accorded probability one are liable to be false.... The ramifications of this approach do admittedly stand in need of further examination. But the position is frankly fallibilistic. Empirical propositions can justifiably be believed and, indeed, admitted into evidence even though it is possible that they are false. (1967, p. 209)

For these reasons, then, we feel that the fact of the fallibility of data poses no threat to the use of the rule of conditionalisation. It is time to move on.

■ j THE PROBLEM OF SUBJECTIVISM

Possibly the most serious of all objections made against the subjective Bayesian theory is that it is simply too subjective. Fisher, in his remark which we quoted in Chapter 3, section **c,** that results concerning the measurement of belief "are useless for scientific purposes", summed up what many thought and still think to be a crucial objection. Science is objective to the extent that the procedures of inference in science are. But if those procedures reflect purely personal beliefs to a greater or lesser extent, as they appear to do if they are constrained only to follow Bayes's Theorem, with no condition other than mere consistency being imposed on the forms of the priors, then the inductive conclusions so generated will also reflect those purely personal opinions. Echoing Fisher, E. T. Jaynes claims that

> the most elementary requirement of consistency demands that two persons with the same relevant prior information should assign the same prior probabilities. Personalistic doctrine makes no attempt to meet this requirement...the notion of personalistic probability belongs to the field of psychology and has no place in applied statistics. Or, to state this more constructively, objectivity requires that a statistical analysis should make

use, not of anybody's personal opinions, but rather the specific factual data on which those opinions are based. (1968, quoted in Rosenkrantz 1977, p. 53)

Alas, neither Jaynes nor his followers are able to live up to his ideal; nor is it possible in principle that they could. No prior distribution reflects only factual data unmixed with anybody's opinions. This is true simply because no prior distribution reflects only factual data, so any given prior distribution will reflect an opinion of some sort. Thus Rosenkrantz, an enthusiastic supporter of Jaynes, defends the uniform prior distributions that tend to arise within Jaynes's theory by pointing out an analogy with current cosmological practice:

Steady-state cosmologists, to take one of myriad instances, start off by assuming the laws of physics are the same in temporally and spatially remote regions of the universe. This, they urge, is surely the simplest assumption. But it is more than that. To assume that different laws obtained a billion years ago would be entirely arbitrary; it would be to import knowledge we do not in fact possess. (Rosenkrantz, 1977, p. 54)

But we don't know that the laws were the same either. Rosenkrantz has failed to see that *any* assumption 'imports knowledge', the assumption that things were essentially the same just as much as the assumption that they were not. So it is with a uniform, or any other prior distribution, as we argued at length (possibly ad nauseam) in Chapter 3. Because it is important, however, let us repeat the fundamental fact once more. *No prior probability or probability-density distribution expresses merely the available factual data; it inevitably expresses some sort of opinion about the possibilities consistent with the data.* Even a uniform prior distribution is defined only relatively to some partition of these possibilities: we can always find another with respect to which the distribution is as biased as you like—or don't like.

Jaynes's objective priors do not exist. But it does not follow, as he, and Fisher, and people too numerous to mention (among them Bayesian personalists) think it follows, that without objective priors the Bayesian theory is constrained to be a record merely of the whims of individual psychology. That quite fallacious inference has been possibly more damaging to rational methodology than any other. We have pressed the analogy with deductive logic, and we shall press it again. Deductive logic is the theory (though it might be more accurate to say 'theories') of deductively valid inferences from premises whose truth-values are exogenously given. Inductive logic—which is how we regard the subjective Bayesian theory—is the theory of inference from some exogenously given data and prior distribution of belief to a posterior distribution. Both logics assign categorical status to certain distinguished types of statement (tautologies, for example, are necessarily true and necessarily have probability 1). Most importantly, as far as the canons of inference are concerned, neither logic allows freedom to individual discretion: both are quite *im*personal and objective.

Moreover the subjective Bayesian theory *does,* as we have seen, incorporate Jaynes's requirement that "two persons with the same relevant prior information" assign the same prior probabilities, but it does so asymptotically, as their data garnered from experience grow without bound. Even then, as we point out in Chapter 11, it characteristically does not take all that much sample data to diminish the different distributions to the point where they are practically identical. Experience is allowed to dominate prior beliefs, in other words, though in a controlled way; disagreement is not eradicated at once, which seems entirely natural, but its effect usually falls off quickly. What more could anybody—reasonably—want?

This consequence of the Bayesian theory, namely the tendency of experience to reduce disagreement, is usually brought out as the sole line of defence against the charge of idiosyncratic subjectivism. While it is important, indeed very important, it is not the sole and should not even be the principal defence, however. For the charge, as we have attempted to show, is quite misconceived. It arises from a widespread failure to see the subjective Bayesian theory for what it is, a theory of inference. And as such, it is unimpeachably objective: though its subject matter, degrees of belief, is subjective, the rules of consistency imposed on them are not at all.

■ k SIMPLICITY

What, though, it may be asked, about invoking a criterion of simplicity as a method of constraining prior distributions—a method which has the virtue both of being objective and of conforming to actual scientific practice? How many times, after all, have we read scientists' claims that it was the great simplicity of such and such a theory which made them repose such high initial confidence in it and remain convinced of its truth long after adverse empirical evidence would have seen off a less intrinsically appealing theory? Why not, therefore, incorporate a criterion of simplicity explicitly as a constraint on ranking prior probabilities?

But it is not at all clear actually how this advice should be followed. Simplicity turns out to be a highly elusive concept, which has so far resisted all attempts to characterise it in any uncontroversial way. Some people maintain that simplicity resides in an organic unity exemplified by the fundamental principles of the simple theory. Others say that it resides in the fewness of the adjustable parameters which the theory introduces. Yet others say it resides in the ease with which computations can be done within the theory. But all these notions, where any clear sense may be made of them, appear to be independent of each other. And it is notoriously difficult, moreover, to make any clear sense of them. Even such an apparently perspicuous notion as paucity of independent parameters is, on inspection, far from ambiguous. Newton's theory, for example, might be thought to possess very few undetermined parameters—some people claim that it contains only one, the gravitational constant. But Newton's theory applied in, say, the kinetic theory of gases, contains of the order, even in the simplest applications, of 10^{23} undetermined parameters, and when further degrees of freedom are added to these ideal models, the number rises proportionately.

The reader must forgive us if we do not add to the vast and inconclusive literature on simplicity. We think that its inconclusive nature is symptomatic of ambiguity and obscurity in the very notion itself. But our main reason for concluding the discussion here is implicit in what we have already said about the nature of our enterprise. We wish to lay down no indefensible a priori principles as universal standards, any more than we should expect a deductive logician to add to the list of logical axioms a statement of the Principle of the Uniformity of Nature, say, simply because it seems to have been adopted implictly or explicitly by so many people. This is not to say that we do not see the Bayesian theory as highly explanatory: we do, but we would take people's apparent sensitivity to considerations of simplicity, in whatever particular form such considerations might take, merely as a true factual statement about one of the components *for them* of a high prior probability.

But here we are obliged to correct a well-known claim of Popper's. According to him, if one takes the paucity-of-parameters analysis of simplicity, then it *"contradicts* the laws of the calculus of probability" to assign greater prior probabilities to the simpler of any two hypotheses (1959, p. 381, his italics). This claim occurs in his discussion of Jeffreys and Wrinch's (1921) so-called simplicity ordering, according to which a pair of rival hypotheses would be assigned probabilities in the way Popper contends to be impossible (*see also* Jeffreys 1961, pp. 45–48 for his further development of the idea of a simplicity ordering). But Popper's claim is incorrect. There is no inconsistency at all in assigning a higher probability to a hypothesis which asserts only that a trajectory is a curve of degree 2 than to that which asserts only that it is of degree 3. Popper's own arguments for his false claim are in fact based on much more than the probability calculus; they are based either on the Classical Theory of probability, or on the principle that the 'more easily testable' a hypothesis is—and one which is simpler, in this particular sense, is more easily testable—the lower should be its prior probability. We have argued at length in Chapter 3 that the use of the Classical Theory of probability to generate prior probability distributions is quite arbitrary. Popper's 'more easily testable = less a priori probable' equation is equally arbitrary: there can be no grounds for assuming that, because fewer independent observations are required to test h_1 than h_2, h_1 is less likely to be true than h_2. There is no more reason to believe this than there is to believe (as Jeffreys did) that h_1 is more likely to be true than h_2 (for a fuller discussion of these issues see Hesse 1974, p. 226–227, and Howson, 1973, 1987, and 1988).

■ I PEOPLE ARE NOT BAYESIANS

In their summary of an influential piece of empirical work, Kahneman and Tversky deliver themselves of the following judgment:

> The view has been expressed . . . that man, by and large, follows the correct Bayesian rule, but fails to appreciate the full impact of evidence [they cite W. Edwards 1968], and is therefore conservative.
>
> The usefulness of the normative Bayesian approach to the analysis and the modeling of subjective probability depends

primarily not on the accuracy of the subjective estimates, but rather on whether the model captures the essential determinants of the judgment process. The research discussed in this paper suggests that it does not.... In his evaluation of evidence, man is apparently not a conservative Bayesian: he is not Bayesian at all. (Kahneman and Tversky [1972] p. 46).

It has been the burden of the foregoing chapters that a Bayesian theory is capable of explaining standard modes of scientific inference where other theories are not. Yet the empirical studies Kahneman and Tversky refer to are taken by these authors to indicate very strongly that people do not use Bayesian reasoning where the Bayesian theory appears to say that they should. Kahneman and Tversky consider the objection to their conclusion, that obtaining posterior probabilities even relative to simple Bernoulli trials requires a degree of practice and sophistication which most people do not possess; they dismiss it, however, citing as evidence that the sorts of departures from the model they find seem to be always of the same type, and independent of the mathematical sophistication of the subjects, and even of their acquaintance or not with basic probability theory. How can we—indeed, can we—continue to regard the Bayesian theory as explanatory in the face of adverse evidence like this?

However, apportioning the blame between a central hypothesis and the various auxiliaries required to test it, when adverse results are obtained, is known to be a problematic affair (this is the Duhem problem, discussed in Chapter 4, section **h**). There is no justification for the definitive conclusion arrived at by Kahneman and Tversky that people are not in agreement with Bayesian precepts in the way they process data. All the cases which are supposed to show this depend on the assumption of the subjects' tacit acceptance of some type or other of probability model which the testers think appropriate. If the subjects had chosen some different model, for whatever reason, then the conclusions to be drawn may be entirely different. Indeed, the author of one of the most celebrated studies to which Kahneman and Tversky refer questions their conclusion for that reason: the subjects' perception of the appropriate model may not have been the testers' (Phillips, p. 1983, p. 531).

However, we should be surprised if on every occasion subjects were apparently to employ impeccable Bayesian reasoning, even in the circumstances that they themselves were to regard Bayesian procedures as canonical. It is, after all, human to err, and sometimes to err in very distinctive and persistent ways. It is instructive to compare the situation described by Kahneman and Tversky with a rather striking and very uniform result (one of the present authors has tested it himself on a group of American freshman and sophomore students) of an experiment, devised by P. C. Wason (1966) to test subjects' performance of, on the face of it, a simple deductive task. Four cards are placed flat on a table. Each card has an integer between 1 and 4 inclusive printed on one face and a letter on the other. The uppermost faces of the cards are

and the subjects are asked to name those cards, and only those cards, which need to be turned over in order to determine whether the statement, 'if a card has a vowel on one side, then it has an even number on the other', is true. Wason discovered that the vast majority of his subjects indicated either the pair of cards E and 4, or only the card 4. The correct answer is, of course, the pair E and 7.

This empirical result has proved to be remarkably persistent:

> Time after time our subjects fall into error. Even some professional logicians have been known to err in an embarrassing fashion, and only the rare individual takes us by surprise and gets it right. It is impossible to predict who he will be. This is all very puzzling... (Wason and Johnson-Laird, 1972, p. 173)

Puzzled Wason and Johnson-Laird may be, but about one thing they are certainly clear: these subjects did get the answer wrong. Moreover, even the subjects themselves eventually agreed on that. Now this observation has an obvious relevance to Kahneman's and Tversky's dramatic claim, made in the light of evidence anaologous to Wason's, that we are not Bayesians. Wason has shown, by this and other empirical studies, that we are not consistently deductive logicians in practice. But he has not shown, nor did he claim to have shown, that we are not deductive logicians in some other important sense. For we ourselves nevertheless constructed those deductive standards and consciously attempt to meet them, even though we sometimes fail, and in some cases nearly always fail. By the same token, it is not prejudicial to the conjecture that *what we ourselves take to be correct inductive reasoning* is Bayesian in character that there should be observable and sometimes systematic deviations from Bayesian precepts.

■ m CONCLUSION

One of the reasons why one expects deductive reasoning to exercise a more-or-less widely felt and obeyed constraint on the way people reason is because it is truth preserving. Probabilistic reasoning also possesses a characteristic which authorises it to exercise no less a regulatory function: its rules, as we observed in Chapter 3, are broken on pain of committing inconsistency. It is, we suggest, for this reason that divergence from the norm set by the probability calculus is also regarded as deviant. Certainly ever since people chose to express their uncertainty in terms of the odds they thought fair, they have felt themselves explicitly constrained by the axioms of the probability calculus, and while it was not until this century that it was explicitly proved that obedience to the calculus is a necessary condition for fairness, there can be little doubt that that result was taken for granted.

The discovery of the probability calculus, together with the usual formula connecting (fair) odds and probabilities, can now be seen to be part of the great scientific renaissance of the seventeenth century. The probability calculus became the foundation of a mathematical theory of uncertainty, of enormous potential scope and power, which simultaneously generated a quantitative logic of inductive inference and bound together the new mathematical concept of probability with another developed at about the same time, utility, to produce a theory of rational action. The mathematicians of the eighteenth century, and to a lesser extent the nineteenth, divided their time between developing the new physics and extending the probability calculus and the theory of inductive inference and rational decision based on it: among these pioneers, Huyghens, James and Daniel Bernoulli, Laplace, and Poisson stand out as pre-eminent.

On the way, however, paradoxes began to appear in the programme, mostly connected with the Principle of Indifference but also—as a criterion of rational action—with the principle of expected utility. These problems, especially those within the theory of probability itself, seemed at one time, in the early years of this century, so intractable that many people, like Fisher and Popper as we have seen, wrote off the account of probability on which the programme was based. But they were wrong: in the middle years of this century, shortly after Fisher and Popper penned their obituaries, secure foundations were finally laid. Von Neumann and Morgenstern put utility theory on a consistent basis, and Ramsey and de Finetti realised that an adequate theory of epistemic probability can dispense with pseudo-objective principles like that of Indifference without giving up its claim to impose quite objective standards of consistency in reasoning involving such probabilities. The probabilities might be personal, but the constraints imposed on them by the condition of consistency are certainly not—a distinction still not widely grasped even today, and whose failure to be appreciated continues to vitiate so much contemporary discussion.

We have written this book in an attempt to convince believers in 'objective' standards in science that there is nothing subjective in the Bayesian theory *as a theory of inference*: its canons of inductive reasoning are quite impartial and objective. We want this simple truth to be more widely appreciated, and not only this one. Equally, we want to demonstrate to those same people that this is the *only* theory which is adequate to the task of placing inductive inference on a sound foundation. The rival claims of the other approaches we have examined in the previous chapters are quite spurious and often do not withstand even a cursory inspection. We hope that we have been at least partially successful in achieving these objectives: the final judgment must, however, as always, be the reader's.

BIBLIOGRAPHY

Altman, D. G., Gore, S. M., Gardner, M. J., and Popcock, S. J. 1983. 'Statistical Guidelines for Contributors to Medical Journals', *British Medical Journal*, vol. 286, 1489–493.

Armendt, B. 1980. 'Is there a Dutch Book Argument for Probability Kinematics?', *Philosophy of Science*, vol. 47, 583–89.

Bacon, F. 1620. *Novum Organum*. In *The Works of Francis Bacon*, vol. 4, J. Spedding, R. L. Ellis, and D. D. Heath, eds. 1857–58, London: Longman and Company.

Babbage, C. 1827. 'Notice respecting some Errors common to many Tables of Logarithms', *Memoirs of the Astronomical Society*, vol. 3, 65–67.

Barnett, V. 1973. *Comparative Statistical Inference*. Chichester: Wiley.

Bayes, T. 1763. 'An essay towards solving a problem in the doctrine of chances', *Philosophical Transactions of the Royal Society,* vol. 53, 370–418. Reprinted with a biographical note by G. A. Barnard in *Biometrika*, 1958, vol. 45, 293–315.

Bernoulli, J. 1713. *Ars Conjectandi*. Basiliae.

Bolzano, B. 1837. *Wissenschaftstheorie* (English translation by Rolf George published under the title *Theory of Science,* Blackwell, 1972).

Bowden, B. V. 1953. 'A Brief History of Computation', in *Faster than Thought,* edited by B. V. Bowden. London: Pitman Publishing.

Campbell, R. and Vinci, T. 1983. 'Novel Confirmation', *British Journal for the Philosophy of Science,* vol. 34, 315–341.

Carnap, R. 1947. 'On the Applications of Inductive Logic', *Philosophy and Phenomenological Research,* vol. 8, 133–148.

_____. 1950. *Logical Foundations of Probability*. Chicago: University of Chicago Press.

_____. 1952. *The Continuum of Inductive Methods*. Chicago: University of Chicago Press.

Carnap, R. and Jeffrey, R., eds. 1971. *Studies in Inductive Logic and Probability*. Berkeley: University of California Press.

Church, A. 1940. 'On the Concept of a Random Sequence', *Bulletin of the American Mathematical Society,* vol. 46, 130–35.

Cochran, W. G. 1952. 'The χ^2 Test of Goodness of Fit', *Annals of Mathematical Statistics,* vol. 23, 315–345.

_____. 1954. 'Some Methods for Strengthening the Common χ^2 Tests', *Biometrics,* vol. 10, 417–451.

Cournot, A. A. 1843. *Exposition de la Théorie des Chances et des Probabilités*. Paris.

Cox, R. T. 1961. *The Algebra of Probable Inference*. Baltimore: The Johns Hopkins Press.

Cramér, H. 1946. *Mathematical Methods of Statistics*. Princeton: Princeton University Press.

Darwin, C. 1868. *The Variation of Animals and Plants under Domestication*. 2 vols. London: John Murray.

Dempster, A. P. 1968. 'Upper and Lower Probabilities Induced by a Multi-valued Mapping', *Annals of Mathematical Statistics*, vol. 38, 325–339.

Dorling, J. 1979. 'Bayesian Personalism, the Methodology of Research Programmes, and Duhem's Problem', *Studies in History and Philosophy of Science*, vol. 10, 177–187.

————. 1982. 'Further Illustrations of the Bayesian Solution of Duhem's Problem'. Unpublished.

Duhem, P. 1905. *The Aim and Structure of Physical Theory* (translated by P. P. Wiener, 1954). Princeton: Princeton University Press.

Dunn, J. M. and Hellman, G. 1986. 'Dualling: A Critique of an Argument of Popper and Miller', *British Journal for the Philosophy of Science*, vol. 37, 220–223.

Edgeworth, F. Y. 1895. 'On some Recent Contributions to the Theory of Statistics', *Journal of the Royal Statistical Society*, vol. 58, 505–515.

Earman, J. 1986. *A Primer on Determinism*. Dordrecht: Reidel.

Edwards, A. W. F. 1972. *Likelihood*. Cambridge: Cambridge University Press.

Edwards, W. 1968. 'Conservatism in Human Information Processing', in *Formal Representation of Human Judgment*, B. Kleinmuntz, ed. 17–52.

Edwards, W., Lindman, H., and Savage, L. J. 1963. 'Bayesian Statistical Inference for Psychological Research', *Psychological Review*, vol. 70, 193–242.

Feiblman, J. K. 1972. *Scientific Method*. The Hague: Martinus Nijhoff.

Feller, W. 1950. *Introduction to Probability Theory and its Applications*, vol. 1. New York: John Wiley.

Feyerabend, P. 1975. *Against Method*. London: New Left Books.

Fine, T. L. 1973. *Theories of Probability*. New York: Academic Press.

Finetti, B. de 1937. 'La prévision; ses lois logiques, ses sources subjectives', *Annales de l'Institut Henri Poincaré*, vol. 7, 1–68. (Reprinted in 1964 in English translation as 'Foresight: its Logical Laws, its Subjective Sources', in *Studies in Subjective Probability*, (edited by H. E. Kyburg, Jr., and H. E. Smokler.) New York: Wiley.

————. 1979. *Theory of Probability*. New York: Wiley.

Fisher, R. A. 1922. 'On the Mathematical Foundations of Theoretical Statistics', *Philosophical Transactions of the Royal Society of London*, vol. A222, 309–368.

————. 1935. 'Statistical Tests', *Nature*, vol. 136, 474.

————. 1936. 'Has Mendel's Work Been Rediscovered?', *Annals of Science*, vol. 1, 115–137.

————. 1947. *The Design of Experiments*, 4th edition. Edinburgh: Oliver and Boyd. (First published 1926.)

————. 1956. *Statistical Methods and Statistical Inference*. Edinburgh: Oliver and Boyd.

————. 1970. *Statistical Methods for Research Workers*, 14th edition. Edinburgh: Oliver and Boyd. (First published 1925.)

Franklin, A. and Howson, C. 1984. 'Why do scientists prefer to vary their experiments?', *Studies in History and Philosophy of Science,* vol. 15, 51–62.

Garber, D. 1983. 'Old Evidence and Logical Omniscience in Bayesian Confirmation Theory', in *Testing Scientific Theories,* edited by J. Earman. Minneapolis: University of Minnesota Press, 99–131.

Giere, R. N. 1973. 'Objective Single Case Probabilities and the Foundations of Statistics', in P. Suppes et al., eds. *Logic, Methodology and Philosophy of Science IV.* North Holland: Amsterdam, 467–483.

———. 1976. 'A Laplacean Formal Semantics for Single Case Propensities', *Journal of Philosophical Logic,* vol. 5, 321–353.

———. 1984. *Understanding Scientific Reasoning.* 2nd edition, New York, London: Holt, Rinehart and Winston.

Gillies, D. A. 1973. *An Objective Theory of Probability.* London: Methuen.

Glymour, C. 1980. *Theory and Evidence.* Princeton, New Jersey: Princeton University Press.

Good, I. J. 1950. *Probability and the Weighing of Evidence.* London: Griffin.

———. 1961. 'The Paradox of Confirmation', *British Journal for the Philosophy of Science,* vol. 11, 63–64.

———. 1962. 'Subjective Probability as the Measure of a Nonmeasurable Set', in *Logic, Methodology and Philosophy of Science, Proceedings of the 1960 International Congress,* edited by E. Nagel, P. Suppes, and A. Tarski. Stanford: Stanford University Press.

———. 1965. *The Estimation of Probabilities.* Cambridge, Mass: M. I. T. Press.

———. 1969. 'Discussion of Bruno de Finetti's Paper "Initial Probabilities: A Prerequisite for any Valid Induction"', *Synthese,* vol. 20, 17–24.

———. 1981. 'Some Logic and History of Hypothesis Testing', in *Philosophical Foundations of Economics,* edited by J. C. Pitt. Dordrecht: Reidel.

Goodman, N., 1954. *Fact, Fiction and Forecast.* London: The Athlone Press.

Gore, S. M. 1981. 'Assessing Clinical Trials—Why Randomize?', *British Medical Journal,* vol. 282, 1958–1960.

Grünbaum, A. 1976. 'Is the Method of Bold Conjectures and Attempted Refutations *Justifiably* the Method of Science', *British Journal for the Philosophy of Science,* vol. 27, 105–136.

Gumbel, E. J. 1952. 'On the Reliability of the Classical Chi-Square Test', *Annals of Mathematical Statistics,* vol. 23, 253–263.

Hacking, I. 1965. *Logic of Statistical Inference.* Cambridge: Cambridge University Press.

———. 1967. 'Slightly More Realistic Personal Probability', *Philosophy of Science,* vol. 34, 311–325.

Haldane, J. B. S. 1945. 'On a Method of Estimating Frequencies', *Biometrika,* vol. 33, 222–225.

Harnett, D. L. 1970. *Introduction to Statistical Methods.* Reading, Massachusetts: Addison-Wesley Publishing Company.

Hays, W. L. 1969. *Statistics.* London: Holt, Rinehart and Winston. (First published 1963.)

Hays, W. L. and Winkler, R. L. 1970. *Statistics: Probability, Inference and Decision,* vol. 1. New York: Holt, Rinehart and Winston, Inc.

Hempel, C. G. 1945. 'Studies in the Logic of Confirmation', *Mind,* vol. 54,

1–26 and 97–121. Reprinted in Hempel, 1965.

———. 1965. *Aspects of Scientific Explanation*. New York: The Free Press.

———. 1966. *Philosophy of Natural Science*. Englewood Cliffs, N.J.: Prentice-Hall.

Hesse, M. 1974. *The Structure of Scientific Inference*. Berkeley: University of California Press.

Hintikka, J. 1965. 'Towards a Theory of Inductive Generalisation', in *Logic, Methodology and Philosophy of Science*, edited by Y. Bar Hillel. Amsterdam: North Holland, 274–288.

———. 1966. 'A Two-Dimensional Continuum of Inductive Methods' in *Aspects of Inductive Logic*, edited by J. Hintikka and P. Suppes. Amsterdam: North Holland, 113–132.

———. 1968. 'Induction by Enumeration and Induction by Elimination', in *Problems in the Philosophy of Science*, edited by I. Lakatos. Amsterdam: North Holland, 191–216.

Hobson, A. 1971. *Concepts in Statistical Mechanics*. New York: Gordon and Breach.

Hodges, J. L., Jr., and Lehmann, E. L. 1970. *Basic Concepts of Probability and Statistics*, 2nd edition. San Francisco: Holden-Day.

Horwich, P. 1982. *Probability and Evidence*. Cambridge: Cambridge University Press.

Howson, C., ed. 1976. *Method and Appraisal in the Physical Sciences*. Cambridge: Cambridge University Press.

Howson, C. 1973. 'Must the Logical Probability of Laws be Zero?', *British Journal for the Philosophy of Science*, vol. 24, 153–163.

———. 1984. 'Bayesianism and Support by Novel Facts', *British Journal for the Philosophy of Science*, vol. 35, 245–251.

———. 1987. 'Popper, Prior Probabilities and Inductive Inference', *British Journal for the Philosophy of Science*, vol. 38, 207–224.

———. 1988. 'On the Consistency of Jeffreys's Simplicity Postulate, and its Role in Bayesian Inference', *The Philosophical Quarterly*, vol. 38, 68–83.

———. 1989. 'On a Recent Objection to Popper's and Miller's Recent "Disproof" of Probabilistic Induction', *Philosophy of Science*, forthcoming.

Howson, C. and Franklin, A. 1986. 'A Bayesian Analysis of Excess Content and the Localisation of Support', *British Journal for the Philosophy of Science*, vol. 36, 425–431.

Howson, C. and Oddie, G. 1979. 'Miller's so-called Paradox of Information', *British Journal for the Philosophy of Science*, vol. 30, 253–261.

Hume, D. 1777. *An Inquiry Concerning Human Understanding*, edited by L. A. Selby-Bigge. Oxford: The Clarendon Press.

Jaynes, E. T. 1968. 'Prior Probabilities', *Institute of Electrical and Electronic Engineers Transactions on Systems Science and Cybernetics*, SSC-4, 227–241.

Jeffrey, R. C. 1970. 'Review of Eight Discussion Notes', *Journal of Symbolic Logic*, vol. 35, 124–127.

———. 1983. *The Logic of Decision*. 2nd edition. Chicago: University of Chicago Press.

———. 1983. *The Logic of Decision*. 2nd edition. Chicago and London: University of Chicago Press.

Jeffreys, H. 1961. *Theory of Probability*, 3rd edition. Oxford: Clarendon Press.

Jeffreys, H. and Wrinch, D. 1921. 'On Certain Fundamental Principles of Scientific Enquiry', *Philosophical Magazine*, vol. 42, 269–298.

Jevons, W. S. 1874. *The Principles of Science*. London: Macmillan and Co.

Kahneman, D. and Tversky, A. 1972. 'Subjective Probability: a Judgment of Representativeness', in *Cognitive Psychology*, vol. 3, 430–454.

Kant, I. 1783. *Prolegomena to any Future Metaphysics*. Edited by L. W. Beck, 1950. Indianapolis: The Bobbs-Merrill Company, Inc.

Kempthorne, O. 1966. 'Some Aspects of Experimental Inference', *Journal of the American Statistical Association*, vol. 61, 11–34.

――――. 1971. 'Probability, Statistics and the Knowledge Business', in *Foundations of Statistical Inference*, edited by V. P. Godambe and D. A. Sprott. Toronto: Holt, Rinehart and Winston of Canada.

――――. 1979. *The Design and Analysis of Experiments*. Huntington, N.Y.: Robert E. Krieger.

Kendall, M. G. and Stuart, A. 1979. *The Advanced Theory of Statistics*, vol. 2, 4th edition. London: Charles Griffin and Company Limited.

――――. 1983. *The Advanced Theory of Statistics*, vol. 3, 4th edition. London: Charles Griffin and Company Limited.

Keynes, J. M. 1921. *A Treatise on Probability*. London: Macmillan.

Kitcher, P. 1985. *Vaulting Ambition*. Cambridge, Mass: MIT Press.

Kolmogorov, A. N. 1950. *Foundations of the Theory of Probability* (Translated from the German of 1933 by N. Morrison.) New York: Chelsea Publishing Company. Page references are to the 1950 edition.

――――. 1965. 'Three Approaches to the Quantitative Definition of Information', *Problemy Peredacii Informacii*, vol. 1, 4–7.

Koopman, B. O. 1940. 'The Bases of Probability', *Bulletin of the American Mathematical Society*, vol. 46, 763–774.

Kries, J. von 1886. *Die Principien der Wahrscheinlichkeitsrechnung. Eine logische Untersuchung*. Freiburg.

Kuhn, T. S. 1970. *The Structure of Scientific Revolutions*, 2nd edition. Chicago: University of Chicago Press. (First published 1962.)

Kuipers, T. A. F. 1978. *Studies in Inductive Probability and Rational Expectation*. Dordrecht: Reidel.

――――. 1980. 'A Survey of Inductive Systems', *Studies in Inductive Logic and Probability*, vol. II, edited by R. C. Jeffrey. Berkeley: University of California Press.

Kyburg, H. E., Jr. 1974. *The Logical Foundations of Statistical Inference*. Dordrecht and Boston: D. Reidel Publishing Company.

――――. 1983. *Epistemology and Inference*. Minneapolis: University of Minnesota Press.

Kyburg, H. E., Jr. and Smokler, E., eds. 1980. *Studies in Subjective Probability*. Huntington, N.Y.: Robert E. Krieger.

Lakatos, I. 1968. 'Criticism and the Methodology of Scientific Research Programmes', *Proceedings of the Aristotelian Society*, vol. 69, 149–186.

――――. 1970. 'Falsification and the Methodology of Scientific Research Programmes', in *Criticism and the Growth of Knowledge*, edited by I. Lakatos and A. Musgrave. Cambridge: Cambridge University Press.

――――. 1974. 'Popper on Demarcation and Induction', in *The Philosophy of Karl Popper*, vol. 2, ch. 5, edited by P. A. Schilpp. La Salle, Illinois: Open Court.

———. 1978. *Philosophical Papers*. 2 vols, edited by J. Worrall and G. Currie. Cambridge: Cambridge University Press.

Laplace, P. S. de. 1820. *Essai Philosophique sur les Probabilités*. Page references are to *Philosophical Essay on Probabilities*, 1951. New York: Dover Publications.

Levi, I. 1967. *Gambling with Truth*. New York: Knopf.

———. 1980. *The Enterprise of Knowledge*. Cambridge, Mass.: MIT Press.

Lewis, C. I. 1946. *An Analysis of Knowledge and Valuation*. La Salle, Illinois: Open Court Publishing Company.

Lewis, D. 1981. 'A Subjectivist's Guide to Objective Chance', in *Studies in Inductive Logic and Probability*, edited by R. C. Jeffrey, pp. 263–293. Berkeley and Los Angeles: University of California Press.

Lindgren, B. W. 1976. *Statistical Theory*, 3rd edition. New York: Macmillan Publishing Co. Inc.

Lindley, D. V. 1957. 'A Statistical Paradox', *Biometrika*, vol. 44, 187–192.

———. 1965. *Introduction to Probability and Statistics, from a Bayesian Viewpoint*. Cambridge: Cambridge University Press.

———. 1970. 'Bayesian Analysis in Regression Problems', in *Bayesian Statistics*, edited by D. L. Meyer and R. O. Collier. Itasca, Illinois: F. E. Peacock.

———. 1971. *Bayesian Statistics, a Review*. Philadelphia: Society for Industrial and Applied Mathematics.

———. 1982. 'Scoring Rules and the Inevitability of Probability', *International Statistical Review*, vol. 50, 1–26.

Lindley, D. V. and Phillips, L. D. 1976. 'Inference for a Bernoulli Process (a Bayesian View)', *The American Statistician*, vol. 30, 112–119.

Lucas, J. R. 1970. *The Concept of Probability*. Oxford: Oxford University Press.

Mackie, J. L. 1963. 'The paradox of confirmation', *British Journal for the Philosophy of Science*, vol. 38, 265–277.

Mallet, J. W. 1880. 'Revision of the Atomic Weight of Aluminum', *Philosophical Transactions*, vol. 171, 1003–1035.

Mann, H. B. and Wald, A. 1942. 'On the Choice of the Number of Intervals in the Application of the Chi-Square Test', *Annals of Mathematical Statistics*, vol. 13, 306–317.

Medawar, P. 1974. 'More Unequal than Others', *New Statesman*, vol. 87, 50–51.

Mellor, D. H. 1971. *The Matter of Chance*. Cambridge: Cambridge University Press.

Mendelson, E. 1964. *Introduction to Mathematical Logic*. New York: Van Nostrand Reinhold Company.

Miller, D. W. 1966. 'A paradox of information', *British Journal for the Philosophy of Science*, vol. 17, 59–61.

Miller, R. 1987. *Bare-faced Messiah*. London: Michael Joseph.

Milne, P. 1986. 'Can there be a Realist Single-case Interpretation of Probability', *Erkenntnis*, vol. 25, 129–132.

———. 1987. 'Physical Probabilities', *Synthese*, vol. 73, 329–359.

Mises, R. von. 1939. *Probability, Statistics and Truth*. Originally published in German in 1928. First English edition, prepared by H. Geiringer. London: George Allen & Unwin.

———. 1957. Second English edition, revised, of *Probability, Statistics and Truth*.

———. 1964. *Mathematical Theory of Probability and Statistics*. New York: Academic Press.

———. 1963–64. *Selected Papers*, edited by Ph. Frank, S. Goldstein, M. Kac, 2 vols. Providence, Rhode Island: American Mathematical Society.

Mood, A. McF. 1950. *Introduction to the Theory of Statistics*. New York: McGraw-Hill Book Company, Inc.

Musgrave, A. 1975. 'Popper and "Diminishing Returns from Repeated Tests"', *Australasian Journal of Philosophy*, vol. 53, 248–253.

Neyman, J. 1937. 'Outline of a Theory of Statistical Estimation Based on the Classical Theory of Probability', *Philosophical Transactions of the Royal Society*, vol. 236A, 333–380. (Page references are to the reprint in Neyman, 1967.)

———. 1941. 'Fiducial Argument and the Theory of Confidence Intervals', *Biometrika*, vol. 32, 128–150. (Page references are to the reprint in Neyman, 1967.)

———. 1952. *Lectures and Conferences on Mathematical Statistics and Probability*, 2nd edition. Washington: U.S. Department of Agriculture.

———. 1967. *A Selection of Early Statistical Papers of J. Neyman*. Cambridge: Cambridge University Press.

Neyman, J. and Pearson, E. S. 1928. 'On the Use and the Interpretation of Certain Test Criteria for Purposes of Statistical Inference', *Biometrika*, vol. 20, 175–240, part I; 263–94, part II.

Neyman, J. and Pearson, E. S. 1933. 'On the Problem of the Most Efficient Tests of Statistical Hypotheses', *Philosophical Transactions of the Royal Society*, vol. 231A, 289–337. (Page references are to the reprint in Neyman and Pearson's *Joint Statistical Papers*, 1967. Cambridge: Cambridge University Press.)

Nickles, T. 1985. 'Beyond Divorce: Current Status of the Discovery Debate', *Philosophy of Science*, vol. 52, 117–207.

Niiniluoto, I. 1983. 'Novel Facts and Bayesianism', *British Journal for the Philosophy of Science*, vol. 34, 375–379.

Pearson, E. S. 1966. 'Some Thoughts on Statistical Inference', in *The Selected Papers of E. S. Pearson* 276–283. Cambridge: Cambridge University Press.

Pearson, K. 1892. *The Grammar of Science*. (Page references are to the edition of 1937, London: J.M. Dent & Sons Ltd.)

Peto, R. 1978. 'Clinical Trial Methodology', in *Proceedings of the International Meeting on Comparative Therapeutic Trials*. Paris: Springer International.

Phillips, L. D. 1973. *Bayesian Statistics for Social Scientists*. London: Nelson.

———. 1983. 'A Theoretical Perspective on Heuristics and Biases in Probabilistic Thinking', in *Analysing and Aiding Decision*, edited by P. C. Humphreys, O. Svenson and A. Vari. Amsterdam: North Holland.

Poincaré, H. 1905, *Science and Hypothesis*, translated by W. J. G. (Page references are to the edition of 1952, New York: Dover Publications.)

Polanyi, M. 1962. *Personal Knowledge*, 2nd edition. London: Routledge and Kegan Paul.

Pollard, W. 1985, *Bayesian Statistics for Evaluation Research: An Introduction*. Beverly Hills: Sage Publications.

Popper, K. R. 1959. 'The Propensity Interpretation of Probability', *British Journal for the Philosophy of Science*, vol. 10, 25–42.

———. 1959a. *The Logic of Scientific Discovery*. London: Hutchinson.

———. 1963. *Conjectures and Refutations*. London: Routledge and Kegan Paul.

Popper, K. R. and Miller, D. W. 1983. 'A proof of the impossibility of inductive probability', *Nature*, vol. 302, 687–88.

Pratt, J. W. 1962. In Birnbaum, A. 'On the Foundations of Statistical Inference', *Journal of the American Statistical Association*, vol. 57, 269–326.

———. 1965. 'Bayesian Interpretation of Standard Inference Statements', *Journal of the Royal Statistical Society*, 27B, 169–203.

Pratt, J. W., Raiffa, H., and Schlaifer, R. *Introduction to Statistical Decision Theory*.

Prout, W. 1815. 'On the Relation Between the Specific Gravities of Bodies in Their Gaseous State and the Weights of Their Atoms', *Annals of Philosophy*, vol. 6, 321–330. (Reprinted in *Alembic Club Reprints*, No. 20, 1932, 25–37. Edinburgh: Oliver and Boyd.)

Ramsey, F. P. 1931. *The Foundations of Mathematics and Other Logical Essays*. London: Routledge and Kegan Paul.

Redhead, M. 1974. 'On Neyman's Paradox and the Theory of Statistical Tests', *British Journal for the Philosophy of Science*, vol. 25, 265–271.

———. 1985. 'On the Impossibility of Inductive Probability', *British Journal for the Philosophy of Science*, vol. 36, 185–191.

———. 1986. 'Novelty and Confirmation', *British Journal for the Philosophy of Science*, vol. 37, 115–118.

Reichenbach, H. 1935. *Wahrscheinlichkeitslehre. Eine Untersuchung uber die logischen und mathematischen Grundlagen der Wahrscheinlichkeitsrechnung*. Leiden. English translation: *The Theory of Probability*. Berkeley: University of California Press, 1949.

Rosenkrantz, R. D. 1977. *Inference, Method and Decision: Towards a Bayesian Philosophy of Science*. Boston: D. Reidel Publishing Co.

Salmon, W. C. 1981. 'Rational Prediction', *British Journal for the Philosophy of Science*, vol. 32, 115–125.

Savage, L. J. 1954. *The Foundations of Statistics*, New York: Wiley.

———. 1962. 'Subjective Probability and Statistical Practice', in *The Foundations of Statistical Inference*, edited by G. A. Barnard and D. R. Cox 9–35. New York: John Wiley and Sons, Inc.

———. 1962a. A prepared contribution to the discussion of Savage, 1962, 88–89, in the same volume.

Schlaifer, R. 1959. *Probability and Statistics for Business Decisions*. New York: McGraw-Hill Book Company, Inc.

Seidenfeld, T. 1979. *Philosophical Problems of Statistical Inference*. Dordrecht: D. Reidel Publishing Company.

Shimony, A. 1970. 'Scientific Inference,' in *Pittsburgh Studies in the Philosophy of Science*, vol. 4, edited by R. G. Colodny. Pittsburgh: Pittsburgh University Press.

Silvey, S. D. 1970. *Statistical Inference*. Baltimore, Maryland: Penguin Books.

Skyrms, B. 1977. *Choice and Chance*. Belmont: Wadsworth Publishing Company.

———. 1984. *Pragmatics and Empiricism*. New Haven: Yale University Press.

Smart, W. M. 1947. 'John Couch Adams and the Discovery of Neptune', *Occasional Notes of the Royal Astronomical Society*, no. 11.

Smith, C. A. B. 1961. 'Consistency in Statistical Inference and Decision', *Journal of the Royal Statistical Society*, Series B, vol. 23, 1–25.

Snedecor, G. W., and Cochran, W. G. 1967. *Statistical Methods*, 6th edition. Ames, Iowa: The Iowa State University Press.

Spielman, S. 1976. 'Exchangeability and the Certainty of Objective Randomness', *Journal of Philosophical Logic*, vol. 5, 399–406.

———. 1977. 'Physical Probability and Bayesian Statistics', *Synthese*, vol. 36, 235–69.

Sprott, W. J. H. 1936. Review of K. Lewin's: *A Dynamical Theory of Personality*, *Mind*, vol. 45, 246–251.

Stas, J. S. 1860. 'Researches on the Mutual Relations of Atomic Weights', *Bulletin de l'Académie Royale de Belgigue*, 208–336. (Reprinted in part in *Alembic Club Reprints*, No. 20, 1932, 41–47. Edinburgh: Oliver and Boyd.)

Stuart, A. 1954. 'Too good to be true', *Applied Statistics*, vol. 3, 29–32.

Suppes, P. 1981. *Logigue du Probable*. Paris: Flammarion.

Swinburne, R. G. 1971. 'The Paradoxes of Confirmation—a Survey', *American Philosophical Quarterly*, vol. 8, 318–329.

Thomson, T. 1818. 'Some Additional Observations on the Weights of the Atoms of Chemical Bodies', *Annals of Philosophy*, vol. 12, 338–350.

Urbach, P. 1981. 'On the Utility of Repeating the 'Same' Experiment', *Australasian Journal of Philosophy*, vol. 59, 151–162.

———. 1985. 'Randomization and the Design of Experiments', *Philosophy of Science*, vol. 52, 256–273.

———. 1987. *Francis Bacon's Philosophy of Science*. La Salle, Illinois: Open Court Publishing Co.

———. 1987a. 'Clinical Trial and Random Error', *New Scientist*, vol. 116, 52–55.

———. 1987b. 'The Scientific Standing of Evolutionary Theories of Society', *The LSE Quarterly* vol. 1, 23–42.

Van Fraassen, B. C. 1980. *The Scientific Image*. Oxford: Clarendon Press.

Velikovsky, I. 1950. *Worlds in Collision*. London: Victor Gollancz Ltd. (Page references are to the 1972 edition, published by Sphere Books Ltd.)

Venn, J. 1866. *The Logic of Chance*. London.

Wason, P. C. 1966. 'Reasoning', in *New Horizons in Psychology*, edited by B. M. Foss. Harmondsworth: Penguin.

Wason, P. C. and Johnson-Laird, P. N. 1972. *Psychology of Reasoning*, London: Batsford.

Watkins, J. W. N. 1985. *Science and Scepticism*. London: Hutchinson and Princeton: Princeton University Press.

———. 1987. 'A New View of Scientific Rationality', in *Rational Change in Science*, edited by J. Pitt and M. Pera. Dordrecht: D. Reidel Publishing Company.

Williams, P. M. 1976. 'Indeterminate Probabilities', in *Formal Methods in the Methodology of Empirical Sciences,* edited by M. Przelecki, K. Szaniawski, and R. Wojcicki. Ossolineum and D. Reidel Publishing Company, 229–246.

———. 1980. 'Bayesian Conditionalisation and the Principle of Minimum Information', *British Journal for the Philosophy of Science,* vol. 31, 131–144.

Zahar, E. 1983. 'Absoluteness and Conspiracy', in *Space, Time and Causality,* edited by R. Swinburne. Boston: Reidel.

INDEX

Note: Because of their constant occurrence, the following terms have not been listed: Bayesians, Bayesianism, and statistics. Bayes's Theorem is listed only once, where its definition is first given.